DESIGNING BABIES

DESIGNING BABIES

BABIES

The Brave New World of Reproductive Technology

Roger Gosden

 W. H. Freeman and Company • New York

For Bob Edwards,
the scientific godfather of many happy families

Text design by Blake Logan

Library of Congress Cataloging-in-Publication Data
Gosden, R. G.
  Designing babies: the brave new world of reproductive technology by
  Roger Gosden.
      p.   cm.
  Includes bibliographical references and index.
  ISBN 0-7167-3299-8 (cloth)
    1. Fertilization in vitro, Human—Moral and ethical aspects.
  2. Fertilization in vitro, Human—Social aspects. 3. Human reproductive
  technology—Moral and ethical aspects. 4. Reproductive technology—
  Moral and ethical aspects. I. Title.
  RG135.G67   1999
  176—dc21   99–17786
                                                                  CIP

Printed in the United States of America

First printing, 1999

W. H. Freeman and Company
41 Madison Avenue, New York, NY 10010
Houndmills, Basingstoke RG21 6XS, England

Contents

Acknowledgments

Writing about human biology and medicine with enough clarity for nonscientists to understand and sufficient accuracy for colleagues to approve is one of the most difficult challenges I have ever undertaken. I am grateful to my publishers for enabling me to put my perspective into print and for encouraging me to express myself freely and the facts as honestly as I can.

I owe many debts of gratitude to people who have assisted me toward my goals, though I alone am responsible for any blunders that remain. Many people—some unknowingly—have given me encouragement or have been providers of information during the preparation of *Designing Babies*. I'd like to acknowledge their unfailing help and record the pleasure of working with them.

The recipe of this book was originally conceived in the offices of my London agent, Maggie Pearlstine. I feel privileged to be in distinguished company on her list of authors and lucky to have the support of an enthusiastic promoter of medical and science writing. When I was writing *Cheating Time*, she asked her assistant, Matthew Bayliss, to be a reader of my manuscript, and he provided lots of helpful feedback. When *Designing Babies* was conceived, he acted as a go-between with publishers. After Matthew's career as an author and scriptwriter was launched, I was twice lucky with his successor, Toby Green. A shared fondness for Darwinia cemented a productive relationship from the start. He has screened and advised on almost every aspect of the manuscript. I am deeply grateful for his perceptive suggestions and for curbing my tendency to wander down the many enticing alleyways leading off my main path.

This is my second book with W. H. Freeman and Company, and I thank Jonathan Cobb and his successors, John Michel and Erika Goldman, for giving me the chance to put my pet subject into print. Despite the ocean between us, they have been very responsive and supportive and have shown a fine understanding of my difficulties in trying to balance a busy professional life with an ambition to write for general readers.

First among my colleagues, it gives me pleasure to thank one of the members of our embryology team, Jenny Krapez. She has not only undertaken background research but also given me plenty of cheerful encouragement when I needed it. My personal assistant, Vivienne Ingham, has been a wonderful help throughout, especially at the editing stage. Somehow she managed to wave a wand over my administrative cauldrons and overheated telephone, freeing me for uninterrupted stretches of concentrated writing. Without Jenny's and Vivienne's help, I doubt that the recipe for this book would have ever reached the cooking stage.

Two people have played important roles in the final stages of the project. Charles Drazin, a writer and publisher by profession, has spent a lot of labor advising me on style and content. He has brought freshness and clarity to the manuscript, lightening my shoulders when other pressures were beginning to bear down. The other helper is my friend Kay Elder, who is an internationally respected specialist in my field and based at Bourn Hall Clinic, Cambridge. She kindly read and made valuable suggestions about the biomedical aspects of the manuscript.

The list of people who have supplied information and advice is long and distinguished, and it is difficult to know where to draw the line. The following deserve special mention: Julia Berryman, Peter Brinsden, Nigel Brown, Elizabeth Bryan, Antonia Byatt, Chris Barratt, David Cove, Howard

Cuckle, James Drife, Robin Dunbar, Gordon Dunstan, Robert Edwards, Malcolm Faddy, Tim Glover, John Harris, Adele Harrison, Lisa Jardine, Matthew Kaufman, Janice Kerr, Henry Leese, Peter Liu, Gerald Mason, David Miller, John Maynard Smith, Randy Nesse, Robert Plomin, Bill Ritchie, Wulf Schiefenhövel, Lynne Selwood, Roger Short, Arlene and Kylie Smyth, Betsi Stephen, John Stephens, and Jenny Uglow.

Because this is a book about reproductive biology and medicine, I am pleased to have this opportunity to thank colleagues in the Centre for Reproduction, Growth and Development at Leeds University. It is a special pleasure to mention the three medical directors of the Assisted Conception Units, Tony Rutherford, Vinay Sharma, and Adam Balen, who have done so much to raise the Centre to prominence as a treatment and research facility for reproductive medicine. And to all the scientists, doctors, and nurses who work in our laboratories and clinics—thanks for being so patient while I was preoccupied during the closing stages of the book.

My final tribute is the most heartfelt. This book has made my wife a writer's widow again, but Carole has borne the role patiently and philosophically and been my sternest and most appreciated critic. Her sacrifice was greater than others', for she had to endure more solitude than she likes and higher electricity bills than she wants. I hope she thinks the product is worth the candle and will enjoy the sight of me padding up and down behind the lawnmower on Saturdays again, like a normal husband. Last, I thank my sons, Matt and Tom. They had no hand in the drafting or editing but remembrance of them was often a source of quiet inspiration when I found the writing hard going. Little did they guess when they saw my head bowed over the keyboard that it was often absorbed with thoughts about them and the awe and significance of bringing a healthy baby into the world.

Prologue

Most people skip the prologue because it is a thing apart and gets between a hungry reader and the main course. I suspect that a literary hors d'oeuvre is served up either to stimulate the appetites of book browsers or to explain the author's obsession. Authors of fiction are driven by the fire of imagination, but the impulses of scientists who set aside their microscopes and test tubes to write about their speciality are different. They are sometimes borne along by a missionary passion to educate or a zealous urge to celebrate nature's wonders. But I felt the urge to write about a controversial field, and this will take even more explanation.

To my enormous good fortune, my career in biology has coincided with a period of unmatched energy and creativity in this science. I was particularly lucky as a graduate student in the 1970s to have worked, almost by chance, in the Cambridge University laboratory where a revolution in reproductive science was underway, as my supervisors pioneered in vitro fertilization (IVF). Research into human embryology had been a no-go area: there was grave reluctance to interfere with the creation of new life, which had always been regarded as sacred. There were plenty of obstacles and knowledge was hard won, but, once the breakthrough had been made and the first test-tube baby arrived in 1978, the floodgates of scientific creativity were opened.

In the past two decades, we have been carried downstream on an ever-widening river of discovery. We have seen fertilization by sperm injection, freeze-banking of embryos, postmenopausal motherhood, and genetic tests for embryos, not to mention transgenic animals and, of course, Dolly the sheep. Discoveries now have a habit of leaping out of the laboratory and into the clinic, and people naturally wonder what will be next.

Reproductive technology is neither the only revolution underway in the biological sciences nor indeed the most momentous. Twenty-five years before the breakthrough in IVF, two scientists burst into a pub in Cambridge one evening and declared to bemused customers that they had discovered "the secret of life." Like the drinkers who turned their noses back into their beer, some scientists were doubtful about the portents of the DNA molecule with the double twist. Few imagined that the genetic code would be cracked so quickly and the door thereby opened to genetic screening and engineering, with extraordinary consequences for reproductive science. Archimedes may have foreseen the implications of his new test for gold when he exclaimed "Eureka," but most great scientists have not anticipated so clearly where their discoveries would lead. Space travel was scarcely imaginable for Galileo staring through his telescope, in vitro fertilization was not anywhere in van Leeuwenhoek's mind as he peered at sperm under his primitive microscope, and nuclear medicine did not materialize in the mind of Marie Curie as she isolated the radioactive element that would eventually kill her. The discovery trail may grow hot or cold, but it rarely runs a predictable course and often leads to consequences that were neither foreseen nor intended.

Who could have imagined that the coils of DNA would be unraveled so easily to lay bare the mysterious genes? Only a few years ago, reading the library of genes in human cells seemed an impossibility, but this feat will soon be achieved and the entire human genome will be on compact disks. Nor had we imagined the raft of technologies and consumer products that molecular biology would yield. The step from reading the genes in animals and plants to manipulating them was made surprisingly quickly, and it was greeted with a mixture of delight and disgust. Biology had become, arguably, the most dynamic and controversial of the sciences as it progressed from observing and theorizing to making and changing. To trace the roots of this new confidence in biology, we must look at its first great revolution.

Biology was originally an innocent pasttime dominated by eccentric naturalists and inoffensive botanists. This tradition was so strong that even I grew up in the 1960s on bird watching and fungus forays, and I knew my flora. One of the great amateurs had as a young man considered careers in both medicine and the church but settled down instead to the peaceful life of a village squire outside London. Before settling down, Charles Darwin set out on a global voyage aboard the *Beagle* in 1831 and returned to England five years later bursting with ideas that would eventually launch a scientific whirlwind. His theory of natural selection shook conventional wisdom just as thoroughly as industry was changing the landscape of England. After years of observations and painstaking experiment, he was able to show that the roots of human genealogy were far older and ran much deeper than had been previously suspected. Even more provocatively, he produced compelling evidence to suggest that cruel competition and chance had dictated our origins.

According to Darwin's theory, natural selection decides which individuals are fit to survive and breed. So powerful was this idea that it quickly engaged not only fellow biologists but also intellectuals who were more bent on human social improvement than studying worms and barnacles. In Europe and America, social Darwinists argued that the unrestrained breeding habits of the lower orders would dilute a society's finest qualities generation by generation. Their remedy for this ill-conceived threat was eugenics, which discriminated in favor of those from middle- or upper-class backgrounds, like themselves, who were judged fit to reproduce. The rest were urged to exercise reproductive restraint. Had Darwin lived to see what had been made of his theory, he would have been dejected. He was a liberal by inclination, and eugenics was inimical to his compassion for the underdog. Yet there was a real dilemma, which the philosopher Bertrand Russell summed up: "The doctrine that all men are born equal, and that the differences between adults are due wholly to education, was incompatible with [Darwin's] emphasis on congenital differences between members of the same species." Science tells us what we can do, not what we ought to do, and not for the first time it had produced a harvest of unintended consequences.

In the aftermath of political purges of physically and mentally disabled people and "aliens" early in the twentieth century, the technological applications of genetics and reproduction became greatly feared. The Nazis had no time for human rights in the hubris of eugenics for shaping national destiny. The political danger may have passed, but these two sciences are still blamed for harming a healthy society and undermining time-honored human relationships, even though the traditional nuclear family has been breaking down for many years. Of all their applications, interference with the process of human procreation itself has been held to be the most baleful—an arrogant pretense at playing God. That is why I chose to begin this book with the fictional story of Victor Frankenstein, who epitomizes our notions of the "evil scientist." Unlike Dr. Frankenstein, who performed his experiments surreptitiously, most scientists work for the public good in the full glare of peer scrutiny and publicity. Yet still, just as he was damned for recreating life ex utero, they have had to endure public horror every time a breakthrough is announced—from the first tentative attempts in the 1960s to fertilize human eggs in vitro to genetic testing of embryos and cloning today. Childless couples may welcome news that brings them a better chance of having a normal baby, but many in the media prefer to conjure up nightmarish images of clones marching in serried ranks out of Huxleyan laboratories or weird embryos growing in artificial wombs. While these images can be faintly amusing, most observers are left perplexed and the researchers hurt at being dehumanized for their search after truth. We cannot be sure that the fruits of science will never be misappropriated, but surely it is self-evident that knowledge must always be better than ignorance.

I think that the contradictory attitudes to science—both fawning over it and fearing it—are intriguing, both sociologically and biologically. Reproduction is something that individuals do and the rest of society cares about. It is perhaps the most meaningful act of our mortal existence, and aspirations to have children are deeply rooted in human psychology. The reproductive sap rises strongly at the due season of our lives and is not easily thwarted.

Today, we are more concerned about the abuse of reproductive freedom than about the vividness of *Brave New World*. Rather than being the cruel instrument of totalitarian government, eugenics has become transformed as people seek to fulfill their ambition to have the most treasured object—a healthy child. Over the centuries, we have secured food supplies, controlled epidemic diseases, and even managed from time to time to achieve political stability. In the twentieth century, couples became able to choose the number and timing of their children, and great advances were made in overcoming infertility. But reproduction is still a chancy business and often fails after an apparently good start or produces a less than perfect product. If the twentieth century was notable for fertility control, in the twenty-first the emphasis of research will switch to producing a baby that is free of defects and attractive and arrives with perfect timing.

The social background against which the powerful technologies of reproduction and genetics are emerging today is freer than before and its citizens have the means to pay for reproductive services. Except in China, where social eugenics is looming again, society by and large does not seek to impose a narrow standard on us. We are now invited as individuals to pick and choose from what is on offer in the reproductive marketplace, according to personal tastes and circumstances. This is the new era of eugenic consumerism.

Another unintended consequence of scientific know-how that originally helped people overcome infertility is that it now helps them decide the kind of baby they want to have.

We have already gone some way down the road toward the "designer baby," even if conventional means of conception and gestation look set to continue for a long time to come. In America (and this will surely spread elsewhere), people choose eggs and sperm from "prize" donors advertising on the Internet, and surrogate wombs are offered for rent. In Asia it is common to try to influence the odds of having a boy or a girl. Many people take steps to avoid possible mental or physical disability: genetic counseling and screening before conception are widely used where there is a history of disease in the family. Most healthy women now accept routine screening and genetic diagnosis to check the condition of the fetus and to have the option to terminate pregnancy if advisable. Where the line is drawn between acceptable and unacceptable defects in a fetus varies enormously. Some pregnancies are clearly regarded as more precious than others, but biology throws little light on how such value-laden decisions are to be made. The more that market forces prevail and patients become aware of the possible choices available, the more ethical dilemmas doctors have to face. They fear that features that were once regarded as trivial blemishes could become grounds for couples to choose abortion and try again. Might they not now object to a fetus with genes that predispose it to developing, say, cancer later in life? Perhaps one day they will be able to determine the color of their children's eyes. Can such a trifling choice be far behind? And what will the attitude be if it becomes possible to clone and engineer the best possible child? Doctors are now on the horns of a dilemma, as the advances of medical science offer reproductive choices that never existed before.

The new danger is that the definition of what constitutes a "good life" will narrow under consumer demand and breed intolerance of individual differences and underachievement. Some parents might be hard-hearted toward their child if he or she does not come up to the expectations technology promised. And what if you can't afford the benefits of this new science? Could the social gulf widen between the children who are "enhanced" and those whose parents were unable to share the dream?

Most researchers and clinicians are wise enough to keep their heads below the parapet of controversy. They are too busy in their labs and clinics to concern themselves with public paranoias. This is a pity because they are in the best position to explain the benefits of their work. Darwin's chief supporter, Thomas Huxley, was no stranger to controversy and was a great communicator of science. He wrote, "Some experience of popular lecturing had convinced me that the necessity of making things plain . . . was one of the very best means of clearing up the obscure corners in one's own mind." When I began this book I aimed to explain whither the road for science and technology of reproduction—which I had joined over a quarter of a century ago. The farther I traveled along that road the more I became as interested in the fears that greeted the scientific breakthroughs as in the discoveries themselves. And the more I reflected on the heavy investment people make in their children, the more I doubted that they make frivolous choices.

These thoughts convinced me that humans have been trying to overcome the limitations and mistakes of nature far longer than assisted reproductive technology or the idea of the "designer baby" have been around. And so this book is not only about the technology of reproduction but about why we care so very much about this area of our experience.

1

The Myth and the Monster

Abominable Creation

At the foot of the Spanish Steps in the heart of Rome stands a pink house three stories high where you can view memorabilia of three great English poets—Shelley, Byron, and Keats. In a small side chamber the portraits of two women stare at each other across the room, as if in an eternal love–hate relationship. The dark-eyed southern beauty was Claire Clairmont, a mistress of Byron—and perhaps also of Percy Bysshe Shelley—and stepsister to Mary, the English rose on the opposite wall. These girls were born into a family of radical intellectuals. Their father was William Godwin, a libertarian and social reformer. His wife, Mary Wollstonecraft, was a friend of Tom Paine and wrote *A Vindication of the Rights of Women* and other essays advocating equal opportunities. She died in 1797 of puerperal fever, eleven days after giving birth to Mary.

Like her mother, Mary displayed an independent spirit from an early age. In 1814, when only 17, she eloped with the 22-year-old Shelley. They married two years later after his first wife, Harriet, drowned herself in the Serpentine in London's Hyde Park.

In the summer of 1816 Shelley and Mary rented a property beside Lake Geneva with Byron and Claire. The weather that summer was wet, and one evening when the conversation was flagging, Byron proposed a new indoor entertainment: "We will each write a ghost story." To encourage the others he began a tale about vampires that evolved in the telling and survives as a fragment in his poem "Mazeppa." We may suppose that Shelley and Claire rose to the challenge too, but their stories have been lost or were never recorded. Mary sheepishly admitted that her mind had gone blank, and she felt even more mortified the following morning when the others teased her. She spent several nights tossing and turning in bed vainly struggling to conjure up an idea "which would speak of the mysterious forces of our nature, and awake thrilling horror." Eventually, the seeds of an idea were sown when she overheard a conversation between Byron and Shelley about some notorious experiments attributed to Dr. Erasmus Darwin, grandfather of the more famous Charles.

In those days, science and technology were discussed in the drawing rooms of every educated family. They had yet to challenge orthodox religion, and it was entirely natural for men of science to mix with men of letters and the cloth. Joseph Priestley, the chemist who discovered oxygen, was also a Unitarian philosopher and a theologian. Sir Humphry Davy, remembered today for his miner's lamp and for discovering metallic elements and laughing gas, also wrote poetry. His close friends, the poets Wordsworth and Coleridge, often attended his scientific lectures and demonstrations. And some years later, when Charles Darwin was vacationing at Freshwater, on the Isle of Wight, Alfred Tennyson would regularly wander down from his house for a conversation.

Darwin's grandfather was a physician and inventive genius who set science to verse; he was also a rebel and philanderer who delighted in scandalizing respectable society. One of the rumors circulating in the chattering classes of early nineteenth-century England, a rumor Mary Shelley had overheard that night, was that he had carried out some successful "experiments in spontaneous generation." In those days it was not beyond the bounds of credibility that living things could materialize spontaneously from inanimate or dead matter. If microbes and maggots could form de novo, why not a human baby? Erasmus Darwin and his friend Anna Seward, so the rumor went, had mingled veal broth and mashed potatoes in a glass vessel "according to art" and created human life, but the baby was "dissolved back into its parent broth" because Anna had impatiently shaken "the gravid bottle"!

The tale was only mischievous gossip, but it set Mary thinking. The following night she was caught up in a vivid dream about science turned into black art: "I saw the pale student of unhallowed arts kneeling beside the thing he had put together. I saw the hideous phantasm of a man stretched out, and then, on the working of some powerful engine, show signs of life, and stir with an uneasy, half vital motion. Frightful it must be, for supremely frightful would be the effect of any human endeavor to match the stupendous mechanism of the Creator of the World." She feverishly put pen to paper the following morning while the memory of the dream was still fresh, and it became the inspiration of her gothic thriller, *Frankenstein*.

The brilliant young scientist Victor Frankenstein had, in a fit of hubris, made a chimeric monster by stitching together bits of cadavers stolen from fresh graves. He put the brain of a professor together with the brawn of a blacksmith and other choice pieces of flesh to make the "perfect" man. When the macabre experiment was complete, he used a spark of electricity to bring the creature to life. The idea that electricity could reanimate a body was a powerful and credible notion in those days because it was still a mysterious new force. The Italian physician Luigi Galvani had shown that touching the muscles of dead frogs with metal rods made them twitch, so it seemed plausible that electrostatic forces might resurrect a corpse. Benjamin Franklin had a lucky escape when he flew a kite in a thundercloud to demonstrate electrical forces, and perhaps Frankenstein was Mary's clever play on Ben's name.

Victor used his inventive genius to try to create a superman. But unlike God, who was pleased after the sixth day of creation, Victor was revolted by the sight of the creature he had made and tried to destroy it. Like Shakespeare's Caliban, it was a tragic figure longing for human company, love, and beauty. Only after it was thwarted and cast out did the monster become aggressive and use his great cunning and strength to pursue a vendetta against society and his "father's" family. In attempting to recreate human life, Victor had plundered nature and had tried to play God. Thus was born the legend of the mad scientist. Mary's story so appealed to her husband that he urged her to embellish and finish what was to become her magnum opus and a touchstone for everyone who fears the power of science.

Mary lived in an era when Britain was becoming a global power and the industrial workshop of the world. Arkwright, Watt, Stephenson, and other inventors were demonstrating the power of water and coal to drive factories and transportation. Erasmus Darwin and Josiah Wedgwood (Charles Darwin's other grandfather) joined together with other learned gentlemen in the newly industrialized Midlands to form a hothouse for new ideas called the Lunar Society; its members were naturally called "Lunaticks." While the pace of progress was quickening at home, explorers from the far fringes of the British Empire brought back exotic specimens and tales of strange peoples and customs that challenged conventional beliefs. It was an era of optimism, discovery, and change, but also of conflict. The Industrial Revolution created a new lower class of people and triggered an epidemic of crime. Machinery was destroyed in Nottingham factories in 1811 when workers protested against the loss of traditional livelihoods. Luddites marched and rioted elsewhere too. But the Establishment had fewer qualms about the growing power of science and industrial technology, for these developments were kindling prosperity.

Hope and Anxiety

Many of the fathers of modern science were ordained in holy orders. They were convinced that their inventions and discoveries were theologically justified, for, as the psalmist had written, "Thou hast given (man) dominion over the works of thy hands; thou hast put all things under his feet." They did not agonize about interfering with nature for they were "thinking God's thoughts after him." They hoped to find signs of his existence and character in the natural world, whether peering through a microscope or scanning the heavens. No less a figure than Francis Bacon had said that knowledge is "a rich storehouse for the glory of the Creator and the relief of man's estate," and "God never wrought a miracle to convince atheism, because his ordinary works convince." Windmills, clocks, ploughs, and other "useful arts" were human rewards for exercising our God-given intelligence. Christian hope and scientific progress flowed in a single stream of history and destiny.

But the time came when every new discovery began to drive a wedge between science and religion. As the power to explain natural phenomena grew, the number of gaps that needed to be plugged by divine explanations diminished, and it seemed to the Victorians that scientific reason would eventually make God redundant altogether. Before the French Revolution, the radical philosopher Diderot had expressed the conflict rather bluntly: "Do you see this egg? With it you can overthrow all the schools of theology, all the churches of the earth." He was not advocating throwing eggs at the cardinal of Paris, but simply suggesting that nature could be a stronger basis of metaphysics than traditional religious beliefs.

The tendency to depend on natural explanation rather than divine interpretation continued to grow, and huge controversy was caused when the geologist Charles Lyell published his *Principles of Geology* in 1830. This gave a very different account of the earth's beginnings from that set out in Genesis and contradicted Archbishop Ussher's estimate of 4004 B.C. for the date of creation. The warts and wrinkles on Earth's surface, Lyell argued, proved that the planet was vastly more ancient; its skin had been transformed by violent volcanoes and gigantic movements and preserved the relics of long dead creatures and plants. Lyell's book enthralled the young Charles Darwin, who read it during his circumnavigation of the world aboard the *Beagle*; it helped to shape his theory of evolution. By the time *On the Origin of Species* was published in 1859, some bitter divisions had opened up between science and the church. Most famously, at the 1860 Oxford meeting of the British Association, Darwin's "bulldog," Thomas Huxley, badly mauled Bishop Wilberforce ("Soapy Sam") for scornful prejudice against the new biology. Even so, evolution did have its disciples among the clergy, like the author Charles Kingsley, but traditional religion seemed to be on the retreat like a tide going out with a "melancholy, long, withdrawing roar," as Matthew Arnold put it.

Science was fast becoming a new establishment with its own priesthood (the professors), institutions, (scientific academies), and creed (the supremacy of reason and truth). It was also developing its own language, which was just as obscure to the uninitiated as the Latin Mass. New adherents were as mesmerized by the fresh insights of the scientific method as Jesus's disciples had been by his teachings 2,000 years earlier. In the world of fiction, Frankenstein's professor at Ingolstadt had proclaimed that the authority of scientists was almost divinely inspired: "They penetrate into the recesses of nature, and show how she works in her hiding-places. They have acquired new and almost unlimited powers and can command the thunder or heaven." Unlike others who "just talked," scientists could "do." The signs were visible and compelling in the cities of industrializing England. Here was the revolution of human hope and progress that Jonathan Swift had expressed a century earlier: "Whoever could make two ears of corn or two blades of grass to grow upon a spot of ground where only one grew before, would deserve better of mankind, and do more essential service to his country than the whole race of politicians put together."

But worries about the impact of the scientific and technological revolution soon began to surface. No longer regarded as innocent interpreters of the divinely inspired hieroglyphs of nature, scientists and inventors were recast as meddlers who might upset the delicate balance of forces that sustain nature and society. The heavy price Frankenstein paid for his hubris was a warning. The prospects of making a completely new living cell or baby from raw chemical ingredients are still remote, although scientists can at least control the reproductive process of existing creatures in ways the Shelleys could never have imagined. But even to discuss the possibilities of genetic manipulation, cloning, and the use of dead bodies or fetuses for therapeutic purposes tends to generate public hostility. While painters, actors and writers present obscene or blasphemous images without reproach, at least in many countries of the West (if not the United States), society feels uneasy when scientists exercise their imaginations too freely. The advance of science has a relentlessness that is absent in the world of the arts, and it is this that fuels anxieties. If Picasso had never been born, no one else would have painted *Les Demoiselles d'Avignon*. But if Watson and Crick had not discovered the DNA double helix and Edwards and Steptoe had not developed IVF, other scientists would have made these breakthroughs sooner or later.

Probably no subject in medical science receives more critical attention from both government and the press than reproductive biology and genetics. Research on human embryos created by in vitro fertilization (IVF) is considered most disturbing of all, even though it is never undertaken lightly and helps us understand why most embryos fail to thrive or are prone to abnormalities. Many countries have forbidden such research. In Switzerland, home of the Frankenstein story, a moratorium was called in 1998 concerning the creation of genetically modified organisms, although fortunately, in the triumph of hope over pessimism, the country voted against a proposal for a total ban. In Britain, the Human Fertilisation and Embryology Act (1990) permits embryo research under license but for only up to 14 days after fertilization. The march of scientific knowledge pauses from time to time, awaiting the discovery of a new theory, technique, or instrument, but it never retreats. Its discoveries can never be destroyed like a canvas that offends or a musical score that grates. Hence the fear that an uncomfortable fact discovered today is bound to be applied sooner or later, possibly for ill.

When *Frankenstein* was published in 1831, it was subtitled "The Modern Prometheus." In Aeschylus's drama, Zeus was enraged when he discovered that Prometheus had stolen fire from heaven and given it to mortals. Prometheus was punished by being chained to a rock, and an eagle was sent every day to devour his liver, which miraculously regrew overnight. Like the tree in the Garden of Eden, knowledge is there for the picking, but a price has to be paid for the fruit. To the Romantic poets, Prometheus was a hero and a civilizing influence who invented useful "arts, though unimagined, yet to be." He represented a scientific ideal: his sacrifice was redemptive of mankind, and in the minds of Byron and Shelley he was definitely on the side of the angels.

> *Thy Godlike crime was to be kind,*
> *To render with thy precepts less*
> *The sum of human wretchedness,*
> *And strengthen man with his own mind.*

The Romantic poets were inspired by the French Revolution, with its promise of a more equitable and prosperous society. But then suddenly the Utopian dream that cut off the ancien régime turned sour. A reign of terror descended on France in the 1790s and was followed by the Napoleonic Wars, the Winter Revolution, and eventually the restoration of the Bourbons. Victor Hugo, more of a realist than the poets, quipped that "revolution transforms everything except the human heart." The Shelleys and other free thinkers began to wonder whether the fruits of knowledge were so appetizing when they could be used by the élite against people as easily as for them.

Nearly two centuries later, the world is transformed by discovery and invention, yet the majority's fear of opening Pandora's box is stronger than ever. The mechanizing of the workplace led to concern about its impact on traditional crafts; developments in nuclear science sparked off fears about radiation, both as a waste product and in weapons of mass destruction, and now many countries are turning their backs on nuclear power generation. The environment is now at the forefront of contemporary concerns and eco-warriors march on the cathedrals of technology. Where once the world was large and hostile, now it suddenly seems rather fragile and vulnerable, and advances in technology are thought as likely to endanger as to benefit it. So it is the scientists who are blamed for the problems, even though knowledge is morally neutral and others profit by their discoveries. Few scientists now bask in the glow of unquestioning public approval, although astronomers, meteorologists, marine biologists, and ecologists are still regarded as relatively benign because they study subjects that are "out there, or down there." Few people lose any sleep about garbage orbiting the planet or fear that radio beacons will reach unfriendly aliens. Perhaps they should.

As architects of the nuclear age, the physicists were once the chief bogeymen. But now biologists have this unenviable distinction. In the first half of the twentieth century biology was considered a poor relation of physics. In the 1920s the famous British geneticist J. B. S. Haldane described his profession in self-deprecating tones: "[The biologist is] a poor little scrubby underpaid man, groping blindly amid the mazes of the ultramicroscopic, engaging in bitter and lifelong quarrels over the nephridia of flat-worms." Esoteric projects and scientific squabbles some-times turn out to be more important than expected, as indeed was Darwin's work on the barnacles, which helped to confirm his evolutionary theory and to destroy the belief that species are immutable.

A less dramatic revolution in biology has already been going on for thousands of years as farmers and animal breeders have manipulated nature by trial and error. One of the products is the yellow wheat field beyond my garden, another my Airedale terrier dozing in front of the hearth. Darwin was well aware of the power of artificial selection to change the characteristics of animals, birds, and plants. He was struck by examples of chickens and doves with extravagant plumage, and cattle with oversized rumps, so beloved of gourmets and farmers yet apparently shunned in nature. Darwin understood the grand scale of things, but the mechanism of inheritance was still unsolved when he died. Had he read a paper by a monk from Moravia, Gregor Mendel, the principles of genetics might have become known in 1865 instead of 1900, when both men were already dead. Mendel's studies of peas in his monastery gardens are fundamental to modern biology because they reveal that heritable factors—or genes—are passed down, usually unchanged, from generation to generation and apportioned to the offspring according to simple rules. But no progress in genetic diagnosis could be made until the chemistry of the gene was known.

A giant leap forward occurred in 1953 when two young scientists at Cambridge University, James Watson and Francis Crick, discovered the structure of DNA. This very long molecule forms the core of the sausage-like chromosomes in the nucleus of cells, which had long been suspected to harbor genes—the inherited information that is held within the nucleus of every cell. Genes are responsible for making the thousands of different types of proteins that make each one of us unique. A few years later the gene code itself was cracked, and a string of new discoveries and techniques quickly followed. As the door to the genetic treasure chest swung open, it became routine to make highly specific, heritable changes first in bacteria and then in plants and animals. These developments have led to products ranging from insulin, growth hormone, and clotting factors to crops that are herbicide resistant or have longer shelf lives. The painstaking work of selecting new strains by plant and animal breeding can now be done quickly and cleanly in the test tube—more like an engineer's work than a farmer's—with implications as profound as the invention of the computer chip.

The next great goal was to read the complete code or genome of a human being. A project was set up in the

1980s in the United States with international collaboration. A budget was approved for about U.S. $1.00 for each letter in the four-letter genetic code (A, C, G, T)—amounting to several billion dollars in total—and a target date set for shortly after the year 2000. At first the going was difficult because the process of reading the code for each gene had to be done manually bit by bit, and the goal for even simple organisms like worms and flies seemed daunting, let alone aiming for the 80,000 genes in linear array on the 22 types of normal chromosomes and 2 sex chromosomes in the human cell. But then automation came along, so the target may be reached sooner than expected. This is biology's moon shot, and we will probably be landing ahead of schedule with plenty of rewarding discoveries made on the journey.

Until the Genome Project was well under way, the method used to find genes was "positional cloning." This method had the advantage that it did not require any prior information about a gene, yet it managed to identify many of the best-known cases, including the genes for Huntington's disease, cystic fibrosis, and breast cancer and the putative "gay gene." It is based on cutting genes with special enzymes to see which members of a family acquire the condition or act as a carrier within a family and comparing the similarities and differences in the DNA.

Molecular biologists would seize the opportunity given by the differences to discover more about the gene. But as the Genome Project advances, a different ploy will be used. By keying into a bioinformatics database, a gene may be traced even if there are only a few consecutive letters of the code available. What is more, technology has become so powerful that no flaw is too small to be detected by genetic diagnosis. Even an elementary mutation in just one cell of an embryo can be identified by amplifying the DNA using a technique called PCR, for which Kary Mullis won the Nobel Prize in Chemistry in1993.

As more genes become accessible for clinical diagnosis and new drugs and gene therapy become available, the doctors of tomorrow will have to be practitioners of molecular as well as conventional medicine. The ability of medicine to cure intractable illnesses has often been exaggerated and its record in prevention has been disappointing, but biomedical research now offers real hope of progress. Imagine going to a doctor's office and having a sample of DNA prepared from your blood and squirted into a machine whirring in the corner. A list of "faulty" genes is quickly displayed on a VDU, and the doctor looks over his spectacles for your reaction. Since we all carry a handful of defective genes that might be passed on to our children or produce disease in ourselves one day, no matter how youthful or unblemished we appear on the surface, we would all be revealed as less than perfect under a gene scanner.

Some people are afraid that our genetic profile might be used against us in the future and that new forms of genetic discrimination will emerge. At the end of a century that has seen so much oppression and prejudice, fears that science has created a new instrument of social control are understandable. Under eugenic laws in Europe and North America in the 1920s and 1930s, the right to reproduce was denied to people judged "unfit" to produce babies of a certain type or "quality." Aldous Huxley cleverly played on our anxieties in his novel *Brave New World*. He depicted a society in which babies were produced on a conveyor belt of in vitro fertilization and cloned to manufacture citizens for designated roles. In the interests of social stability, individual rights were sacrificed and the possessive feelings of parents diluted to promote public spirit. Huxley offered an uncannily accurate prediction of scientific advances, and if in today's more aggressively libertarian society many old anxieties about the use of reproduction for social control have diminished, *Brave New World* can still evoke a deep unease in people as they consider the sinister turns this frightening new technology might take. Hollywood recently took genetic prejudice as the theme of the movie *GATTACA*. The hero manages to overcome the disadvantages of not being genetically enhanced, which is presumably intended to reassure us.

Chief pundit and opponent of "genome intrusion" in America Jeremy Rifkin believes that the "concern over a reemergence of eugenics is well-founded but misplaced. While professional ethicists watch out the front door for telltale signs of a resurrection of the Nazi nightmare, eugenics doctrine has quietly slipped in the back door and is already stealthily at work reorganizing the ethical priorities of the human household." Rifkin is concerned that technologies that begin with the worthy aim of alleviating inherited illnesses, such as cystic fibrosis, will lead to the elimination of fetuses for trivial differences, such as left-handedness or color blindness, as soon as we understand the genetic causes. On the other hand, there are some people who want their children to share the same disability they have inherited, such as deafness or short stature. These seemingly contradictory desires demonstrate how powerful and complex are our private concerns for our children and the difficulty of deciding public responsibility. Setting those aside for now, let us consider what reproductive technology has achieved and where it began.

Instruments of Creation

John Hunter was born in Scotland in 1728 in relatively humble circumstances and rose to become a celebrated anatomist and surgeon to King George III. Although more inclined to scientific pursuits, he practiced medicine for a living and was sought after by patients who had despaired of conventional medicine and quackery. A young couple visited him one day, hoping desperately for a cure for their childlessness. The husband's penis had hypospadias, a condition in which an abnormal orifice prevented the man from effectively inseminating his wife. Hunter advised the couple to collect his semen in a warm cup and inject the fluid into her vagina using a syringe. After a few attempts following Hunter's instructions in the privacy of their own home, the wife conceived and bore a child. This triumph should have been a cause for celebration, but Hunter was wary of publicity and feared censure because he had encouraged the "unnatural" act of masturbation. The secret was kept until after his death when his cousin mentioned the artificial insemination experiment as an aside in an obscure paper describing the anatomy of a hermaphrodite dog.

Ignorance and superstition about infertility were the norms in prudish Victorian society that forbade sex education and research. Children's curiosity about their origins was brushed aside with diversions and anecdotes about birds and bees, storks, and gooseberry bushes: consequently children learned from one another—and often a lot of nonsense. The world of academia was not much more enlightened, and fertility research remained a stagnant backwater for a long time. Even such fundamental facts as the time in the month when a woman ovulates were unknown until the 1930s, and family planning methods were correspondingly ineffective. Medical practitioners could offer little help to people wanting to turn the fertility tap on or off and, feeling helpless, tended to avoid such questions. In 1930s Britain, one woman wrote to women's rights campaigner Marie Stopes, "We had a talk with our own doctor but he appeared unwilling to speak about it, so we did not pursue the matter further." Children were still regarded as "gifts," and, though they arrived at frequent intervals in some families, unluckier couples had meekly to accept their infertility and deny instincts and expectations that others assumed to be their birthright.

After shocking the British public with the candor of her book *Married Love*, published at the end of World War I, Marie Stopes delivered a series of boisterous lectures and founded clinics up and down the country promoting sexual health and family planning. Here, at last, was an advocate for couples who were suffering from an excess or want of reproduction and who dearly hoped to bring their fruitfulness under control. Once the social inhibitions began to lift and more effective methods of fertility control were available, couples started to choose when to have children according to their preferences and circumstances, rather than leaving matters to nature.

The ambition to have a family is still strong in most people, despite the falling birth rate, and it first emerges in our nursery days when we play mothers and fathers. When our make-believe days are over, there are social pressures on top of the old biological urges to try the real thing. Would-be grandparents, aunts, and uncles welcome family additions and can be relied on to drop a few hints to an older couple "before it is too late." Everyone seems to be proud and gain status when a planned child arrives, and, in return, parents hope to win some reciprocal affection from their offspring and some security in old age. Perhaps those who have been denied children recognize the urge better than those who take their fertility for granted. Indeed, the strength of this urge has a biblical authority. God commanded Noah to "go forth and multiply," and Noah begat three sons; many Christian church wedding services still beseech the happy couple to be "fruitful."

On top of all these factors, there is something ineffable about the longing to be someone's parent. After self-preservation and satisfying hunger this is the most fundamental urge, and obviously it is vital for the species. Even the asexual amoeba has to split in two or face extinction. We seldom think this way, but reproduction is the only bid we can make for a small slice of immortality—through the survival of our genes, that is. Whatever heights of achievement we attain in our careers, whatever public acclaim we enjoy, most of us will be forgotten in a couple of generations. But our genes endure in our descendants, even though each set is shuffled and only half of them are passed down to each son and daughter. The "pack" has hardly changed at all over eons of time, and the urge to perpetuate the genes is powerful and universal and wells up from our Stone Age past.

The first timid step taken by science to treat infertility was carried out by John Hunter, but it took another 200 years before a major stride was made. If pressed to state the time and place of this breakthrough, I would say that it came just after midnight on July 25, 1978, in a small hospital in Oldham, Lancashire. After years of trying, the Brown family produced Louise, the first baby conceived by in vitro fertilization. She was born after a decade of research endeavor by the gynecologist Patrick Steptoe and the Cambridge physiologist Robert Edwards, and she emerged into a world full of curiosity about her.

The two men had to conquer not only the difficult problem of conception and growing embryos in a culture fluid, but also fierce opposition, even from their medical colleagues. It was feared, even by some fellow professionals, that interfering with a natural process would produce monsters, and the Frankenstein story was duly trotted out in the newspapers. Signaling their doubts, the much-respected *New York Times* ran the headline "Brave New Baby," and *Time* magazine called the Cambridge laboratory "Orwell's baby farm," implying something unsavory about laboratory conception. Rather than rejoicing at the birth of a healthy baby girl to parents who had made enormous sacrifices, the daring experiment was condemned, and it was feared that a Pandora's box of "human hatcheries" had been opened.

In the years since the breakthrough, assisted reproductive technology (ART) has evolved into a conventional treatment. Nearly 0.2% of American and 1% of British babies are being born after IVF, which is just one form of ART. There are now more than 300,000 IVF babies worldwide. Patients stand in line for treatment, regardless of the stress, discomfort, and risks and despite the fact that the success rate for the treatment is seldom better than 1 in 5. The lengths to which some people will go to have the child they desperately want reveal more clearly than anything else how powerfully we are driven by our desire to carry forward our own kind.

Reproductive technology did not stand still for long before moving forward with another innovation. Originally developed to bypass blocked fallopian tubes so that the embryo can reach the womb, IVF has turned out to be useful for treating all sorts of problems, in males as well as in females. New methods are constantly being invented to help the process of conception. Spare embryos can be frozen at minimal cost, and, since the first little "frosties"—as the first babies who were born from frozen embryos were known—arrived in the mid-1980s, most clinics now offer this service. Egg donation was developed for women whose ovaries are prematurely barren, and it has turned out to be effective for women of more mature years. Biopsies from embryos can be genetically screened for heritable diseases before transferring the embryos to the womb to ensure a healthy baby and avoiding the decision to abort a defective fetus later on. A chance discovery in a laboratory led to another revolutionary treatment (not cure) for men with low sperm counts. Injecting a sperm directly into an egg, a technique called intracytoplasmic sperm injection (ICSI), enables even the most underendowed male to father a child, and it has swept through IVF centers to become standard practice. Such lightning progress causes people to wonder how far the technology will have gone in a few more years' time.

Whereas our parents' generation left matters to nature, we take fertility for granted. Few people of any age need now be involuntarily infertile. If the time-honored methods fail, the problem usually has a technological solution. Consequently, we have come to regard parenthood as more of a right than a privilege and a right that can be exercised at our behest and in the circumstances of our choosing. The majority of family units in the West are still of the nuclear variety—Mom, Dad, and the kids, like the Brown family—but immense social changes are underway. In vitro fertilization, gamete donation (egg or sperm), and surrogacy open the door to previously undreamed-of choices, including solo or homosexual parenthood as well as motherhood past

the age of 50. These are just some of the unintended consequences of scientific advances, and there will be many
more in the future.

Such possibilities provoke strong reactions, and as a
result, reproductive medicine in some countries has
become hedged about with more legislation and regulation
than any other branch of medicine. The main concern has
been to safeguard the interests of children, although surprisingly few problems have emerged so far. Attitudes are
likely to harden as ART moves on from infertility treatment
to more effective prenatal screening and selection and
then, possibly, to genetic enhancement. As we strive to
make our lives as painless and secure as possible, it is natural to want to remove the uncertainties about pregnancy
and our offspring, too.

Reproduction has always been a risky business, and
many embryos perish naturally in the womb before the
woman even realizes she is pregnant. Some reproductive
waste can be reduced by improving health during pregnancy,
but little can be done, as yet, about the large numbers of
poor-quality embryos that are the cause of many a miscarriage. No other mammal we know of is as prodigal with its
seed or has a higher frequency of birth defects. Most of the
prenatal losses are "merciful" because the embryos are
chromosomally abnormal, and any large imbalance in the
number of genes always has some harmful effects. Why the
germ cells are so much more susceptible to abnormalities
than the rest of the body is a mystery, but usually only the
fittest embryos stand a good chance of going the whole dis-
tance of pregnancy. Nevertheless, occasionally an abnormal
embryo is strong enough to escape the selection process
and is born with a defect.

Before the advent of prenatal screening there was no
reliable way of telling whether a baby would be normal or
not. Pregnancy screening and fetal diagnosis now give a
mother information to help her decide whether to take further tests or even have a medical termination. As technology
advances, more and more information about the baby will

become available. The prospect of screening the entire genome at the embryo stage is not very far off. But the closer we look, the more "flaws" we are likely to find. What should our reaction be? Should we help the natural screening process that eliminates most defective fetuses as miscarriages, and, if so, who should decide? Must we not acknowledge that none of us is perfect and wonder who would meet the strictest criteria? It is not always easy to define abnormality, but there may sometimes be value in it. The Norwegian expressionist Edvard Munch, who painted *The Scream* and suffered from mental illness, once said, "I would not cast off my illness, for there is much in my art that I owe to it." The future holds the possibility of correcting genetic faults one day so that fewer fetuses need be aborted. Correcting an inherited disease and avoiding suffering is life-affirming and something that many people would welcome, but enhancing the quality of perfectly healthy embryos smacks of playing God and conjures up the specter of the Frankenstein monster.

We are becoming familiar with genetic modification in producing cereals, vegetables, and fruits that are more resistant to disease and herbicides, produce heavier and more uniform crops, and have longer shelf lives as a result. How far these developments will go depends as much on the attitudes of consumers as on the creativity of science or the faith of investors in the new products. Ideological battle lines have been drawn between those who delight in progress and those who reject it, whether they believe that "nature knows best" or fear encouraging an invasive and "masculine" technology. Even the Prince of Wales has joined the fray and asked, "[Do] we have the right to experiment with, and commercialize, the building blocks of life?"

Imagine the call to arms when a technological equivalent to the one that gave us transgenic wheat and soybeans is introduced into the marketplace of human reproduction. Imagine a superman or a superwoman engineered from a perfect sperm and a perfect egg and carbon copied as identical clones. Will our increasing ability to control the quality as well as the amount of reproduction in a free and competitive society allow a monster to spring loose? Or will nature always try to preserve variety and refuse to sanction such a triumphal takeover by technology? We cannot deny the powerful drive within us to invest our very best in our children, and apply the benefits of discovery. Reproduction may be a private matter, but how far society will be content to leave the choices up to couples and individuals we have yet to see.

2

The Precious Child

The *K* Club

The Reverend Robert Burton was a seventeenth-century don at Christ Church, Oxford. He is mainly remembered as the author of *The Anatomy of Melancholy*, which is full of Jacobean wisdom on every medical subject from ague to impotence. Two centuries later, his book was still widely read, and Byron recommended it as "useful to a man who wishes to acquire a reputation of being well-read, with the least trouble." Burton was an amateur psychiatrist who was fascinated by "melancholia," or depressive illness, an affliction that seems to have been just as widespread in his day as it is in ours. For a celibate clergyman, he also had an uncommon interest in the sex drive and the parenting instinct. It is self-evident that animals have a strong urge to reproduce their kind, for otherwise they would have become extinct long ago. But not all show much responsibility after their babies are born or are willing to make a large personal sacrifice or risk their own lives for the sake of their offspring. Burton mused that the more familiar species are often the exceptions and the most passionate parents: "Nature binds all creatures to love their young ones; an hen to preserve her brood will run upon a lion, an hind will fight with a bull, a sow with a bear, a silly sheep with a fox."

One spring day, while browsing in Burton's *Melancholy*, I became distracted by a little drama going in my garden. Perched atop the apple tree, an agitated female blackbird was clack-clacking angrily and making repeated dives into the newly green beech hedge. My curiosity aroused, I went outside to investigate the rumpus. A thieving magpie was sitting in the hedge just a few inches above the blackbird's nest—a neat cup of dried grass and twigs bearing four green and speckled eggs. Burton might well have witnessed a similar scene in the gardens of Christ Church and added blackbirds to his list of brave parents. After clapping my hands, the disgruntled magpie flew away and, satisfied with my good deed for the day, I returned to my desk. But the next morning the nest was empty.

It is unusual for a member of an alien species to want to shelter or protect another. A she-wolf is supposed to have mothered Romulus and Remus, but that is only a legend. Animals wisely look to their self-interests and rarely look after the young of others unless there is a quid pro quo. Interest in another species is normally parasitical—for the eating thereof or for shelter or a piggy-back ride. Humans are different because we sentimentally indulge the underdog, fawn over our pets, and befriend many other creatures with one or two pairs of legs, but seldom more.

Not one of the blackbird's own kind had come to her aid. Neighboring blackbirds are bitter rivals and strut up and down the lawn in a confrontational manner, like members of rival street gangs. From time to time they raise their wings in anger and, feinting a charge, try to encroach on the invisible line that sets their territories apart. Were I to make an equivalent gesture to my neighbors, it would be considered rude, but posturing is very necessary in blackbird society if pairs are to protect an area large enough to feed a hungry brood. Reproduction is a selfish activity, and we should not expect to find many tokens of altruism unless it is in the interests of both parties to cooperate. On one occasion I did witness an alliance, when a sparrow hawk landed in the garden. Both pairs of blackbirds forgot their differences momentarily and made such a fuss that the raptor went on its way.

From our nursery days, we have read tales about furry or feathered families whose love and sacrifice for their own children (if not for others) knows no bounds. Those who nurture the best we admire the most, perhaps because they represent heroic figures that we idealize in our own species. Yet in nature, parental devotion is not a matter of choice or honor but of necessity. The ovaries and testes play a hormonal symphony to the brain, which then instructs the muscles and glands to respond appropriately—first in courtship, then in mating and incubating, and finally in caring for the babies until they are old enough to look after themselves. Only when the reproductive orchestra stops playing do parents turn their attention to the hardships of their own survival, especially when the inclement season approaches. Once the breeding season is past, resources spent in nurturing the young are diverted to laying down fat and feathers and boosting the immune system to keep parasites at bay.

The blackbird's chief prey, earthworms, also have a keen interest in reproducing their kind. Charles Darwin built a vivarium so that he could study their feeding and breeding habits. Experimenting with a system of flashing lights, he found that their sexual drive was "strong enough to overcome . . . their dread of light." There is a lot of passion on the lawn after darkness, and not just the vermiform variety. Beetles and insects that make their home between the grass blades and roots are prodigious breeders, if poor rearers.

Although by human standards the earthworm's dereliction of their duty to care for their offspring seems feckless, in their world it makes perfect sense. These species usually grow quickly to adult size for a big bang of reproduction before their short lives end, and they shed lots of eggs with few thoughts for the consequences. As long as one or two survive they have not lived in vain. It pays for them to produce lots of babies as quickly as possible, because they live in risky habitats and have many enemies. They have neither time nor need for the finer points of parental responsibility.

Most mammals and birds live in more secure circumstances and can afford to breed at a more leisurely pace. They live longer and yet still assure their line by producing enough offspring to make grandchildren. In reproduction, strategy is everything. The animal that breeds quickly cannot afford to give lots of attention to each baby; conversely, the slow breeder cannot risk neglecting its young. Generally speaking, the slower the rate of breeding the greater the parental devotion. Blackbirds produce two or three clutches of three to five eggs per year for an average life span of only three to four years, and most other garden birds have a similar rate of fertility. But longer-lived species take their time finding a mate and invest much more effort in each chick. Royal albatrosses wander the southern oceans for five years before choosing a permanent mate and making a nest. They lay only one large egg every other year, and both parents travel thousands of miles back and forth to feeding grounds to bring fish and squid to their huge baby until it can join them in the air.

In the mammal and bird worlds there is a spectrum of care. At one end there is the Australian marsupial mouse *Antechinus stuartii*, which indulges in a frenzy of courtship and copulation after its first year, the males then promptly dying from fatigue and immune failure. Their widows are left the task of raising the pups before dying themselves. At the other end of the spectrum, elephants are far less amorous and females come into sexual heat infrequently, but they devote themselves to their calves for the several years they require to reach maturity. Ecologists call animals at this end of the spectrum the *K*-strategists in distinction from the *r*-strategists at the other end. These uncolorful terms have been extracted from an equation that is used to describe the relative constancy (K) of populations in larger, longer-lived species, like elephants and albatrosses, compared with the more variable rate (r) of increase in others, such as antechinuses, worms, and insects.

Humans are at the extreme end of this species spectrum. No other species postpones fertility longer or invests more parental care when it finally begins to reproduce. Our population growth rises ever upward, more like that of *r*-strategist rodents with their population booms and crashes, but we are otherwise archetypal members of the *K* club. The large brain of the human has evolved to provide a greater capacity for learning and complex socially cooperative behavior than any other species. What is more, we do not lose interest in our offspring as soon as they have grown up but usually cherish the bond throughout life. The role of the human family has become uniquely extended by social evolution beyond the biological end of producing heirs and successors, and this has undoubtedly helped to make us such successful apes.

It is in the interests of members of the *K* club to make the finest-quality offspring so that their parenting efforts are worthwhile. Notwithstanding the depths of parental and social depravity to which members of our own species occasionally descend and the desperate straits that sometimes force people to give up their children, we are remarkably devoted to our sons and daughters. Our attitudes to and the value we place on children are defined by the relative scarcity of children, as the anthropologist, Margaret Mead, once pointed out: "Our humanity depends on our relative infertility, upon the long period of human gestation and dependency possible only where there are few children who can be reared long and lovingly." There is, of course, a risk that time, labor, and emotion will be invested in a disappointing "product"—the child born with a serious disability, one who dies prematurely, one who turns to crime or against its parents. Although other animals are undoubtedly sorrowful at the loss of a pup or a calf or a chick, their feelings are probably transient, but being a parent in the *K* club can be heart-rending. If technology promises to deliver more security in the lottery of reproduction, people will surely be tempted to try to avoid the pains and pitfalls and make the child that fulfills a dream. Far from being shocked that people should want to make "designer babies," we should expect them to.

Humans are intensely social creatures, and our behavior sometimes denies our biological roots. There are feckless parents as well as devoted one and societies that exploit children as well as others that are child-centered. Not everyone wants a child, let alone a designer one. Some people regard childbirth as a defining moment in their existence, others seem indifferent to being genetic dead ends without issue. Never in the history of humanity have more people chosen to remain childless, which, in biological terms, is a puzzling attitude for any animal to have. Fears about a dangerous, unstable world and the cult of the individual both erode the reproductive instinct. But although sometimes the urge to beget offspring goes underground, I doubt whether it is ever completely buried. It is not easy to penetrate the froth and social heterogeneity of modern life to discover the emotions and urges that made us successful breeders in the past. But they are never more tellingly revealed than in the reactions of those who are denied children when they want them or those who lose them.

Grieving Rachel

In the early years of marriage, the Old Testament figure Rachel was barren and wanted a child so desperately that she cried out to God: "Give me a child or I die." The hurt was greater because her husband, Jacob, had already proved himself with her less attractive sister, Leah. In desperation, she provided him with her handmaid, Bilhar, to beget children, which was a custom that her grandmother-in-law, Sarah, had employed so that Abraham could father offspring. Later, Rachel was lucky with Jacob and conceived Joseph after wearing mandrakes, or "love apples," which are members of the nightshade family and reputed to have aphrodisiac properties. Rachel regained her poise and happiness as she enjoyed the status and pride that parenthood brings.

So solemn was the requirement to avoid extinction of the family line that the ancient Hebrews were obliged under Levirate law to beget children for their dead brothers. God slew Onan when he "spilled the semen on the ground"—in other words performed coitus interruptus—with his widowed sister-in-law. But on the whole it was the woman who was always blamed for the failure to produce a baby, and this belief is still used as an excuse for abusing and divorcing wives in some pronatalist countries where raising heirs for husbands is a defining female role. Men are often reluctant to admit that they are the cause of a sterile union, and the Patriarchs in the Bible were never held responsible for the "shame" of infertility. Producing lots of sons was necessary if the tribe was to survive, and the misogynist prophet Hosea prayed that Israel's enemies should be defeated by having their instruments of reproduction destroyed: "Give them a miscarrying womb and dry breasts."

Anyone who has had involvement in a fertility clinic knows that the earnest desire to make babies is by no means a thing of the past, but it is at least no longer a matter of shame or embarrassment. Stories about both the misery of infertility and the encouraging advances in research are reported almost daily in newspapers and magazines. In the past the infertile had to bear their suffering silently, but nowadays they can talk openly, and most problems can be overcome, if at a price.

Infertility is high in the league of common medical conditions. Some surveys have shown that up to one in six couples are involuntarily childless at some stage in their lives, and the cause of the infertility rests with the man as often as it does with the woman. Estimates of the problem vary and are expressed differently, although there is little reason to suspect significant national differences. In the United States, the National Survey of Family Growth estimated in 1995 that 6.2 million Americans in the 15 to 44 age group were affected, and 2.7 million of them were sufficiently concerned about infertility to seek medical advice or assistance.

Many people still prefer to keep these matters to themselves, but Americans are generally in the vanguard of openness. At Georgetown University in Washington, D.C., Elizabeth Stephen has surveyed the attitudes of childless couples undergoing infertility treatment and found an intensity of feeling that surprises those who take their own fertility for granted. Like most middle-class Americans seeking treatment, the majority had given little thought to having a family when they were younger and busily building their lives. By their mid-thirties, their biological clocks were running down and hopes of starting a family surfaced, just as conception was beginning to become more difficult. The men in the survey were less perturbed than their wives at the prospect of life without a family, though they admitted that they were "disappointed" when they discovered the problem. The women frequently said they were "devastated." As a man I must protest that my love for my sons cannot be any less than their mother's, but there is a different emphasis between the genders in love and marriage. Byron tried to capture it in words that probably would anger modern feminists:

> Man's love is of man's life a thing apart,
> 'Tis woman's whole existence.

But even feminist writers sometimes agree that the reproductive drive—so often a handicap to female independence—can be all-consuming. Lee Silver at Princeton University comments that for "the vast majority of people . . . the desire to have children is so powerful that it outshines everything else they might possibly want to do during their lives." One of the women in Elizabeth Stephen's survey, called Joan, felt the need for motherhood so strongly that she prepared for sex with a degree of seriousness usually reserved for business. She monitored her urinary dipsticks to check when she was ovulating, and on red days her husband, Dan, canceled his engagements to try with her for a baby. Such a methodical approach may pay off, but the fervor for fertility can diminish the passion of love and the spontaneity of sex. Unfulfilled hopes can strengthen bonds between a couple, but it can also drive them apart.

Most people assume they are fertile unless proved otherwise. Many mistakenly believe that youth, beauty, and vitality are good omens for their reproductive prowess. The discovery of infertility then comes as a shock. Perhaps it is more so to people who have achieved all their other goals—a graduate education, a comfortable home, and a rewarding career. Such people find it hard to understand why they cannot conceive and have to turn to professional help. The trials of reproduction sometimes deflate egos, as one woman in the study admitted: "Ken and I are both goal oriented. We have done very well in what we set out to do, but we have failed at this so far. My husband has a Ph.D. from Stanford, but we can't make this happen." For a Christian couple, Eunice and Roy, the experience of failing to conceive shook their faith: "I can remember sitting in church crying, Why? Why us? If you are really up there, God, why are you doing this to us?"

The experience of having to negotiate fertility hurdles and to pay dearly for what was presumed to be free can affect attitudes to children when they eventually arrive. The natural response is thankfulness, but this is hugely accentuated for those who have been trying to conceive for years. The walls of most fertility clinics are full of portraits of babies and birthday parties sent by adoring parents who have not forgotten the difficulties they had to overcome or the clinical staff who helped and supported them through the ordeal. Births are celebrated all the more if they are hard won, especially if the first success might be the last. When Claire, another woman in the survey, saw a mother smack her child in a shop, she declared that she "could never hit a child like that . . . because this child is a gift." Far from carrying any stigma, the child is likely to be regarded as a special, even miraculous gift. In the Bible, boys born to parents who had suffered the anguish of infertility usually went on to great fame and achievement—Isaac, Joseph, Samuel, John the Baptist. The greater the investment and the more marvelous the conception, the greater the expectations. Parental anticipation is both an opportunity and a liability for children who have been conceived after much fervent prayer or investment in technology. Will children conceived by assisted conception technology be more accomplished when they grow up because of the fact that their parents had difficulties conceiving them?

The yearning to conceive is sometimes driven by a desire to bring happiness to a partner or parent, although this can add to emotional strains in a relationship. An older man taking a younger wife may have already fathered children and be uninterested in more, but she may want them to cement the marriage bond and overthrow jealousy about the rest of his family. An older female may wish to give her younger partner a child he may not want but dares not deny her. That the psychology of human relationships is complex hardly needs stating. Tensions can occur between generations, as well as between partners, when expectations are not fulfilled. Another woman in Stephen's survey had a mother of 78 who was in frail health and wanted a grandchild as soon as was respectable after her daughter had married. The failure to produce a child strained the relationship between mother and daughter, and the daughter became depressed about her own mortality. "I wanted to have someone else to love and to create a new person. It made me think that when my mother dies I won't have a child to replace her. It has made me afraid of growing old alone."

The desire to bear a child can become obsessive, and the costs of infertility treatment are often heavy. Infertility patients are willing to accept considerable discomfort as they undergo a roller coaster of emotions and medical procedures that would be considered humiliating or even dangerous in other circumstances. When IVF was being pioneered, one of Patrick Steptoe's patients volunteered to submit to over 20 treatment attempts. Some patients are all too willing to subject themselves to experimental treatments that stand only a tiny chance of success. Successful treatment does not necessarily extinguish the urge to have another child, as there is then a desire to produce a playmate for the first. If parents are willing to face so many trials and obstacles simply for the chance of having a child, what lengths would they go to or what sacrifices would they make to guarantee its good health and prospects?

Couples who do not succeed after trying again and again, as well as those who are averse to technology, find solace in other outlets. They may take more interest in the children of their relatives and friends or plunge deeper into an absorbing career or pastime. Some opt for adoption, though the numbers of children available have plummeted since the abortion laws were liberalized, and in Britain this former standby is denied to older couples. In America the problems of adoption have been compounded by court rulings in which adoption orders were reversed in favor of biological parents after child custody had been transferred.

If the inability to conceive is due to a total lack of either eggs or sperm, donation of one of these gamete cells provides a good chance of success. Most people have a strong preference for their own biological child and consider donor gametes, if at all, only when other avenues are closed. Others are more accepting of foreign genes as substitutes for their own or the use of a surrogate woman if the female partner has had a hysterectomy, whether she has retained her own ovary with eggs for IVF or has to rely on the surrogate for both eggs and womb. The introduction of a third party to facilitate fertility has attracted widespread condemnation and is still illegal in many countries. The impact of these techniques on both partners and the child deserve more research, although they are difficult to study because of the confidentiality that is usually involved. One woman in Elizabeth Stephen's survey who successfully gave birth after egg donation was not completely satisfied. She was "still grieving that I can't get pregnant like everyone else. I just want to be pregnant again and enjoy it, or maybe I just want my DNA passed on?"

Public convictions about the transcendence of genes—not always well-informed and often exaggerated—are in danger of increasing anxiety about the health of children and the psychological need for parents to have a child biologically related to them. In accepting gamete donation, the role of genetic parenting is surrendered by one partner to please the other and produce a child who will be loved by both. As yet, we cannot be sure whether this need to rely on a third party affects relations between the couple or bonding with their children, but we must hope that this biosocial innovation creates families with as much chance of success and happiness as conventional ones.

It is natural that the difference between male and female roles in reproduction should be reflected in men's and women's attitudes to egg and sperm donation. Men may be reluctant to admit feeling uneasy about sperm donation out of respect to their partners' feelings, but from time immemorial, there has been that old male anxiety "Is it my baby?" Perhaps women are more comfortable with the thought of egg donation because they have always known where the egg came from and were usually sure who the father was. It will be interesting to see whether research will prove or deny these predictions.

Blood ties are still valued and respected in reproduction and revenge. It is said that "blood is thicker than water," and the closer the kin the thicker the relationship. Fifty years ago, when J. B. S. Haldane was asked if he would sacrifice his life for his brother, he cleverly quipped, "No, but I would for two brothers—or eight cousins." According to kinship theory, our interest in others is in proportion to the number of genes shared—50% with siblings and 12.5% with cousins. Haldane was already persuaded of the existence of selfish genes before the term was coined by Oxford biologist Richard Dawkins, though the Bible gave a graphic example long ago. Two prostitutes who lived together were brought before Solomon to judge which of them was the natural mother of a male child. The women had delivered sons at the same time, but when one of the babies died, both had claimed the survivor. Solomon was a good psychologist, if a hard judge, and asked for a sword to bisect the boy so that the mothers could take an equal share. This test of their resolve produced a yielding response from the rightful mother who was willing to give up her share to save the child's life, "for her heart yearned for her son." The other woman declared, "It shall be neither mine nor yours; divide it."

A mother begins to bond with her child even before its birth. The knowledge that a life is growing inside her has a profound effect on her maternal feelings, as do the hormones that bombard her brain. In animals, the sight and smell of the baby during the first few minutes after birth are critical for bonding. If a kid is taken away at birth from a nanny goat, she may reject it if it is returned to her later. Human responses are more flexible, but once the bond is formed it is probably for life. No wonder it is so painful for parents when the time comes for their children to leave home.

In a life that is uncertain from beginning to end, few days are more nerve-wracking than the one when a woman goes into labor. This is the riskiest day in her baby's life and used to carry a heavy toll of maternal mortality too. The medical profession urges women to deliver in a hospital and discourages home births for, although animals, including humans, naturally prefer a quiet, stress-free place at this time, when something goes wrong in labor being on your own can be disastrous. A difficult labor, fetal asphyxia, or maternal hemorrhage require expert medical help—and quickly.

Many pregnancies do not make it to a full-term delivery. Miscarriage is responsible for the loss of about one in six pregnancies, or even more in older women. Some babies are stillborn and others are born too premature to survive long (though the impulse of many parents is to urge doctors to save premature babies' lives, even if the chances are slim and the risk of disability high). Generally speaking, the more advanced the pregnancy, the greater the mother's sense of loss. If bonding has already begun and she regards her unborn child as an individual, it is little comfort for a grieving mother to be told that miscarriage is nature's way of eliminating a blighted embryo or fetus. Fortunately rare, but more painful still, is the double calamity of being diagnosed with cancer during pregnancy. If chemotherapy is needed before the fetus can be safely delivered by cesarean section, there is a high risk of it being seriously harmed and a termination of pregnancy is advisable.

As every parent knows, anxieties do not disappear once a baby has been safely delivered: they are replaced by others. I remember putting my head in our sons' cribs each night before retiring to bed to confirm that they were still breathing, even though there was no reason to be alarmed. Infant deaths sometimes come without any warning, as the name Sudden Infant Death Syndrome (SIDS) implies. Despite an enormous amount of research and a fall in the death rate, SIDS still claims 8 children in every 10,000 annually. The risks are reduced after 4 months of age and are lessened if a baby is laid on its back in its crib and has nonsmoking parents.

Teenage children have a lower annual risk of dying than any other age group, but no parents can protect their child from all dangers, however hard they may try. The loss of a child at this age is perhaps the most grievous. When the parents think their child has survived the most vulnerable stages of life and is set for a full and healthy span of years, a tragic accident or illness can rob them of their most cherished investment. This is a blow from which some people never recover, and it cannot be fully understood except by those who have experienced it. The death of Charles Darwin's daughter, Annie, at the age of 10 in 1851, broke his heart and his belief in the Christian creed. Reproduction is fraught with tragedies as well as joys, and we sorely yearn to make the whole process as risk-free and painless as possible.

What Price the Priceless Child?

As a species, we invest "long and lovingly" in our children, and some people go to extreme lengths to achieve their desire for a family. But those planning a family ought to pause to consider the costs, and many an exasperated parent of teenagers or young adults has had moments of wondering about the wisdom of their investment. The sums laid out for their child's upkeep, education, and recreation, which could have been spent enjoying a personal luxury such as a new car or a time-share apartment in a vacation resort, are indicators of serious personal sacrifices made in the interests of improving their child's health, happiness, and success. The biologist in me puts the sacrifice down to the hormonal "sap" within us and to the gilt-edged asset of genetic capital. But the drive to invest so heavily in our offspring impresses economists and sociologists too.

The American sociologist Robert Shoen posed the question "Why do Americans still want children?" He reached the conclusion that children represent a kind of social capital for families. From this perspective, any desire to make designer babies is not so much a biological phenomenon as a social epiphenomenon. As evolutionary biologists have regarded "wealth" as a proxy for fertility, social scientists see the relationship the other way around. Throughout most of history, high fertility has been socially advantageous to families.

In the past, and still in many parts of the developing world, the arrival of babies was welcomed without fuss and accepted as part of the natural course of events. Additional hands—especially male ones—were always needed on the farm, and a "quiverful of sons" promised more security in old age. It sounds crass to call children commodities, but in the past they had an undoubted economic value that could be quantified, and they often still do.

Nevertheless, society has often been hard on children who are from disadvantaged backgrounds, like the street urchins of Victorian times. The history of child exploitation and abuse led reformers like the Earl of Shaftesbury, aided by Charles Dickens, to introduce laws that protected the interests of children. The Climbing Boys Act (1840) abolished the employment of children as chimney sweeps, and the Coal Mines Act (1842) forbade apprenticeship in the mines for any child under 13 years old. Parliament and pulpit were in accord in condemning child labor, children were gradually removed from the workforce, and by late Victorian times their employment was said to "touch profanely a sacred thing." A rudimentary education, if not more, became compulsory and the minimum age for leaving school gradually rose.

Children spent more and more time at home or at school than in the workplace, but the changes occurred slowly, and it was only with hindsight that the extent of this domestic revolution was appreciated. Not everyone applauded the changes. In 1904, a sour correspondent wrote to *Harper's Weekly*, "We work for our children, plan for them, spend money on them, buy life insurance for their protection, and some of us even save money for them. This last tribute is the most affecting of all . . . saving for our children's start in life . . . is evidence of serious self-denial. We would be better off putting the money in the bank for our old age. . . . Profound must be the depths of affection that will induce a man to save money for others to spend." Suddenly, people woke up to the changes and realized that children had become "useless" and expensive.

Social scientists writing in the national press love to shock parents with the staggering costs of rearing children, and strong indeed must be the procreative urge not to be daunted by such figures. In 1998, a survey by *U.S. News & World Report* predicted that the price tag for raising one child over the next 21 years will average $761,871 for a family earning less than $35,000 a year before taxes, $1,455,581 for middle-income earners, and $2,783,268 for higher earners. The figures include the opportunity costs of foregone income as well as college education, which varies enormously. Few parents could contemplate such a big investment if they were to have as many children as their grandparents; at no time in history has so much been spent on so few. If the large figures spent by wealthy families on their children are extraordinary, low earners nonetheless provide a better indication of the investment people are prepared to make because the sacrifices they make for their children are so much more costly to themselves. In surveys, low-income mothers have often declared that they have sacrificed clothes, entertainment, and vacations for themselves so that their children's basic needs could be met.

The cost of infertility treatment is another indication of how much importance people attach to children. Public provision of infertility services varies from full costs in Israel to patchy subsidies in Britain and a wholly private system in the United States. A single cycle of treatment in a British clinic costs £2,000 to £3,000 or more, depending on the amount and type of drugs used. In the United States, where drugs and professional fees are higher, the equivalent treatment costs around $10,000. Most patients are unlucky first time around and have to try again or switch to a different technique, such as ICSI or egg donation or even preimplantation genetic diagnosis, all of which bear additional costs. These charges add up to a tidy sum for a service that does not guarantee a product; a child born after ART can easily figure as the second most expensive purchase made after buying a home. Yet people still clamor to try these techniques. Those who cannot afford them can only live in hope that research progress will reduce the price tag of ART by developing a more streamlined technology that is less reliant on expensive drugs.

Attitudes to children are contradictory, and perhaps never more so than today. Not so long ago George Bernard Shaw espoused the view that "perhaps the greatest social service that can be rendered by anybody to the country and to mankind is to bring up a family." Would that everyone had such a generous regard for the children of others. Parents reach deep into their pockets for their own children but are more reluctant when called upon to help others. There are contradictions between the private and public value of children. The writer Germaine Greer considers modern society to be "profoundly hostile to children." Schools and hospitals in the public sector that have served the community well in the past are now struggling for resources as we become increasingly reluctant to pay taxes for the education and health of other people's children. W. Norton Grubb argues that "the saccharine myth of America as a child-centered society, whose children are its most precious natural resources, has in practice been falsified by our hostility to other people's children and our unwillingness to support them." The helping hand of paternalism is out of fashion; it has become respectable to look after one's own and pay scant heed to others.

Less Is More

According to a sixteenth-century English rhyme that we learned by heart on our mother's knee:

> *There was an old woman who lived in a shoe,*
> *She had so many children, she knew not what to do.*

In an era when Western women declare that one, two, or at most three children is the ideal family size, anything larger is eye-catching. How the old woman from the rhyme shoehorned her brood into a cramped Elizabethan home beggars our imagination, but in those days the birth of 10 to 20 babies to one woman was not remarkable. In the absence of effective contraception—short of abstinence—babies kept coming one after another until menopause, which was probably welcomed with relief by the women who survived that long.

The overcrowded and unsanitary conditions of the past, combined with a poor diet, encouraged the spread of smallpox, plague, diphtheria, scarlet fever, and tuberculosis in European towns, and the Grim Reaper regularly made his rounds. Both private and public attitudes to children were more offhand then, and relationships were more remote to lessen the grief of a final parting. It was said to be folly to invest much emotional capital in ephemeral beings, and there would always be another child on the way to take the place of any who died. Babies were sometimes baptized with the same name as the lately departed or given the name of another in the family since it was unlikely that both would survive. In rural France, deceased infants were often interred in the backyard, rather as we bury domestic pets today, and without much more ceremony, by the middle and upper classes. The humanitarian French essayist Michel de Montaigne wrote, "I have lost two or three children in infancy, not without regret, but without great sorrow."

Such attitudes may have always prevailed when the chances of child survival were slim. Prehistoric painters in France and Spain never depicted children in their art, though they drew plenty of bison and aurocks. Transitory lives are less likely to be commemorated, though the memory of a child may be permanently imprinted on the body in some societies. While I was staying with the Dani tribe in Irian Jaya (in New Guinea), I sat beside an ancient-looking grandmother and her daughter, both in rags, and the son-in-law, who was traditionally dressed in only a cap of feathers and a penis gourd. All the grandmother's fingers had been cut off at the first joint, and when I asked my interpreter how this had happened, I was told that she had used a special stone axe to chop off one digit for every child she had lost. Most of the older tribal women had missing fingers (never thumbs). The fathers, in contrast, sliced off part of their ear with a parang to leave their fingers intact for pulling bowstrings. Quite apart from the demands of their society for a public exhibition of grief, there was no doubt that the loss of each child, marked by these mutilations, was heartfelt. It may seem odd that people who only recently emerged from the Stone Age should make a greater display of feeling than did families during the late Middle Ages in Europe; perhaps it is because their families are not so large as were families in the Middle Ages.

The social historian Lawrence Stone found that there were few public expressions of mourning for children in England between 1500 and 1800, and few parents attended their children's funerals. As in France, even babies from the gentry were sometimes buried like paupers, "sewn into shrouds made of cheap sacking and thrown into big, common graves." Even within living memory, in the little village where I live in West Yorkshire, many children were buried without a memorial. There are 13 unmarked graves on the east side of the graveyard, farthest from the church. I remember an "owd fella" recalling that when his younger brother died in infancy, his father carried the box down to the churchyard unceremoniously after darkness. Tucked into the wall of another corner is one of the few tombstones commemorating a baby: "Mabel Walker—not given, only lent."

In America, a greater show of public grief for children started at the turn of the twentieth century, when mortality rates were falling, though under-5s still accounted for 40% of all deaths. The waste of young lives, including street kids, from accidents involving electric street cars in New York City disturbed people so much that May Day 1926 was declared "No Accident Day" to draw attention to the hundreds of girls and boys who had been killed. The Reverend Cuyler, a well-known clergyman in the city, wrote a little book entitled *The Empty Crib* as a literary memorial to his son, Georgie, who had died suddenly after an illness. In previous generations, such sentimental treatment of child death would have been unthinkable, but he received thousands of sympathetic letters from similarly bereaved readers. Funerals for children became more somber—and more expensive, too. The coffins were made of good-quality wood with a brass name plate, lock, and key. In some parts of Europe, it became popular for better-off parents to adorn tombs with little angels or to erect a statue of their deceased child. And if dead children were being lavished with pomp and ceremony, the living ones were also being cherished more. Victorian parents in London had to pay "baby farmers" for fostering services, and neglected children fell into the hands of unscrupulous Fagins. By the 1930s in Europe and America there had been such a turnaround that unwanted babies were in demand for adoption, and the charges were as high as $1,000 and more.

Improvements in public and personal hygiene and in diet were enabling more children to survive the hazardous years to puberty when the chances of long life were much better. Children may no longer have been an economic asset in the modern society that was emerging, but they were at least considered deserving of proper medical attention. This in turn had the effect of further raising rates of survival.

Hippocrates had set a trend over 2,000 years before by advising people to pay attention to external sores and rashes and teething problems in minors but ignoring more serious matters of internal medicine. This relative inattentiveness persisted until the nineteenth century when child patients were still being treated, if at all, as miniature adults with the same approach to physiology and pathology. Pediatrics emerged as a medical discipline only when the fathers of the French Revolution, wanting healthy citizens and soldiery, founded l'Hôpital des Enfants Malades in Paris in 1802. For the next 50 years France led the world in medical education and research into both adult and child maladies, recognizing the distinction of age and maturity at last. Across the English Channel, public provision lagged behind. Not until 1852 was the Hospital for Sick Children established in Great Ormond Street by Charles West, with the help of Dickens and Shaftesbury. In the following years other hospitals dedicated to sick children were founded in Edinburgh, Liverpool, and elsewhere, though there were plenty of arguments among the medical consultants whether pediatrics should be regarded as a specialty in its own right. This proposal received less resistance in America, where the first children's hospital was opened in Philadelphia in 1855. The American Pediatric Society was established in 1889, and by 1912, a Bureau of Child Health had been established. American pediatrics remains preeminent to this day.

At the same time the common childhood killer diseases—smallpox, diphtheria, and so on—were being vanquished, family sizes were in decline. It was as if the population were swinging even farther to the *K*-end of the reproductive spectrum, although the factors responsible were social and environmental rather than changes in human biology or genetics. It made economic sense to have fewer children when there was a good chance that they would all survive. As family sizes contracted and prosperity increased, children became more and more objects of devotion as parents and grandparents no longer had to divide their time and resources among so many.

Nowhere are changes in parent–child relationships more dramatic than in modern China, where the "little emperors" reign. Under 1979 legislation, rural Chinese were allowed a quota of only two children and urban couples only one. Those who break the law have to pay a penalty of 50,000 yuan ($6,000) and are harassed by officials. This policy, intended to keep the lid on the Republic's population of 1.2 billion, is one of the biggest experiments of social engineering ever attempted by a government. There is an old Chinese saying that is now seldom heard: "More children, more happiness." Today's society is being turned upside down as an older generation becomes the majority members of the family, with the grandparents sharing a parenting role with mother and father. This is called the four-two-one family. On Saturday afternoons in large Chinese cities, McDonalds and KFC outlets are crowded with demanding and overweight youngsters. It is difficult to resist spoiling your one and only child or grandchild, especially if you had a deprived childhood yourself, as so many of today's parents in China did during the experiment that Mao Zedong set loose on the nation—the "Cultural Revolution." And in an increasingly competitive society, there is intense parental anxiety and pressure for children to succeed, with top priority given to education and foreign language skills. Schools have even had to put up signs forbidding parents to attend the lessons themselves so that they can help their child with homework at the end of the day. Parental sacrifice is common as mothers and fathers seek to fulfill through their child the thwarted ambitions of their own youth. The one-child policy in China encapsulates in the space of just one generation the many social changes happening at a slower rate elsewhere.

Higher standards of living, more leisure time, and fragmentation of communities have coincided with smaller families. Parents are now focusing attention on just one or two very valuable children. They are putting all their eggs into one or two baskets, so to speak, to produce the ultimate K-selected child, though by social rather than biological evolution. After laying careful plans for childbearing and rearing, anything less than a perfect outcome is disappointing, and the birth of a disabled child greatly feared. If reproductive technology and genetics were able to ease parents' fears about the intelligence, health, and longevity of their offspring and even satisfy ambitions for a talented child, would they pay for the privilege? Parents may know deep down that the search for the perfect baby is illusory, but at what point does boundless love close the door to a technology that just might make the difference between celebration and calamity?

This brings me to the nub of another question. Why does reproduction fail so often and sometimes produce a child that is less than perfect? Why didn't evolution weed out the weak and unsuccessful models along the way so that humans could sit above animals as confidently in their reproductive prowess as in their intelligence? The grand highway of evolution has never been a respecter of individuals in the way that we care for our own sons and daughters. It pays regard only to the numbers of healthy offspring that survive to breed and assure the genetic line. This conclusion is so obvious in what we observe in the natural world but so hard to accept for ourselves and our loved ones.

3

The Pursuit of Perfection

Well-Born

We might think that the high value the Victorians placed on child health and survival was wholly commendable, but there was also a sinister aspect. As well as a growth of children's charities, hospitals, and schools, there was a growing obsession with the "quality" of babies and fears about a weakening national bloodstock. In every generation there are those who venerate the "good old days," resist change, and decry modernity. But even progressive thinkers were saying that children were less fit and intelligent than before and that Britain was consequently slipping into decline.

Victorian Britain boasted an Empire on which the sun never set and that depended on vigorous administration and military firmness. It was a country that had been used to winning, but, toward the end of the century, it began to suffer a number of humiliating setbacks with nationalist struggles in the Indian subcontinent and the African colonies. In the Sudan, the Mahdi had boxed in General Gordon and taken Khartoum; in South Africa, the Zulu nation, armed under Cetshwayo only with spears, inflicted several humiliating defeats on British troops. When the Boer Wars began, the nation was lamenting the fact that as many as one in three of the men who were screened for military service had to be rejected as medically unfit. The decimation of Britain's youth in World War I compounded the national pessimism, and it was feared that with only the unfit left for the girls back home to marry, the next generation would be even feebler than the last.

These anxieties were given a gloss of scientific respectability by reference to Darwinian theory. Forty years after the death of the great biologist, evolution by natural selection was widely accepted and carried intellectual gravitas. Charles Darwin had shown in *On the Origin of Species* that each generation of animals produces more offspring than needed to maintain the population so that competition for food and breeding partners caused a "struggle for existence." Only individuals that were best fitted to the conditions would survive. Darwin realized that our species had shrugged off natural selection but hesitated to apply any conclusions from nature to human society, voicing his apprehensions only occasionally: "We civilized men . . . do our utmost to check the process of elimination; we institute poor-laws; and our medical men exert their utmost skill to save the life of every one to the last moment. . . . Thus the weak members of civilized societies propagate their kind. No one who has attended to the breeding of domestic animals will doubt that this must be highly injurious to the race of man." He mainly kept his head below the parapet on these matters, but his younger cousin, Sir Francis Galton, had no misgivings about standing up and airing his opinions.

Galton was a polymath who took in his stride pioneering research on weather maps, fingerprint identification, twinning, and quantitative biology. Today, he is mainly remembered for studying human inheritance and for coining a word that would haunt the next century—"eugenics," from the Greek meaning "well-born." Galton regarded eugenics as a corollary to evolutionary theory, for "natural selection rests upon excessive production and wholesale destruction; eugenics on bringing no more individuals into the world than can properly be cared for, *and those only of the best stock.*" Rooted in evolutionary theory, eugenics sounded intellectually respectable and, with its goal of improving the well-being of the population, seemed socially responsible, too. But the eugenics movement, which started in biology, ended up—to its lasting shame—as ideology.

Believing that cream should be permitted to rise to the top, Galton encouraged families to enter competitions for the prettiest and brightest-looking baby. He also studied the pedigrees of famous men to discover the secret of high achievers and, in his book *Hereditary Genius* reached the conclusion that "if a twentieth part of the cost and pains were spent in measures for the improvement of the human race that is spent on the improvement of the breed of horses and cattle, what a galaxy of genius might we not create." On a darker and more practical note, he advocated "the hindrance of marriages and the production of offspring by the exceptionally unfit"—what we now call "negative eugenics." He urged people to act responsibly in matters of reproduction, as in everything else, but it was not long before others were suggesting compulsory sterilization and abortion for people too feckless or ignorant to care about the public interest. Had Galton been able to envision the future course of science, he would have applauded the genetic screening of parents and fetuses and the creation of sperm banks for superintelligent donors. The repair of genetic faults and "enhancement" of babies—"positive eugenics"—were undreamed of in his day, but they became the ultimate goal for his successors, who urged that technology should seek to effect a destiny that evolutionary selection was always too blind and slow to reach.

In the 1600s—long before eugenics and Darwinism—the Reverend Robert Burton of Oxford expressed similar thoughts: "It is the greatest part of our felicity to be well born, and it were happy for humankind if only such parents as are sound of body and mind should be suffered to marry." Making babies was never regarded as just a private matter; it was held to be a function of the social organism. Burton went on in a pessimistic vein to say that "in giving way for all to marry that will, too much liberty and indulgence in tolerating all sorts, there is a vast confusion of hereditary diseases, no family secure, no man free from some grievous infirmity." Blame for the ills of society has often been laid at the door of vulnerable or minority groups, and long before the advent of "scientific" eugenics it was tempting to suggest that relieving them of their reproductive liberty was the remedy. Historical precedents seemed to point to the selection of the fittest as a formula for a successful society. I remember being taught at school that asceticism and social responsibility were responsible for the military glory of Sparta: parents were forced to abandon weak boys at birth, and when the survivors reached a certain age, they were thrown into a pit full of water to let nature decide who was fit enough to become a citizen of that supposedly virtuous state. Galton and his successors, who received plenty of instruction in the classics, no doubt impressed by the achievements of ancient societies, were full of hope for what science could do for theirs.

After Galton died in 1911 and World War I was over, others carried forward the eugenics banner, and in England none with greater passion than Marie Stopes, the redoubtable family planning crusader. She traveled up and down the country setting up women's health clinics, appealing to working women for reproductive restraint, and berating the dysgenic effects of war. She declared that it was in women's interests to have no more than two or three children and worried that those who demonstrated most enthusiasm for birth control were the educated elite and gentry.

Concerns about being overwhelmed by inferior stock soon spread from Britain to continental Europe and America, where eugenics was vigorously taken up and introduced into the academic curriculum of many universities. Fears were fanned that the "original" American stock was being diluted by mass immigration of aliens from southern European countries. Bowing to trade union pressure and pandering to public prejudice, President Coolidge enacted legislation in 1924 to restrict the numbers entering the country. In referring to people with learning difficulties and physical disabilities, the National Conference on Race Betterment had menacingly warned a few years previously that "in prolonging lives of defectives we are tampering with the functioning of the social kidneys." The climate was emerging for eugenic laws in many states to authorize compulsory sterilization and abortion, and thousands of U.S.

citizens went under the knife over the two decades after World War I. People who were judged ill-equipped mentally or physically to produce and rear children were the targets; they were more often females than males. No matter how convincingly the opponents argued against social Darwinism—and some came from the nation's top universities—rhetoric won the day.

In Britain, a national review committee was established in the 1930s, and although eugenic legislation failed to make it into the statute books, poor and illegitimate children were still being sent into isolation and virtual servitude in Australia right up to the 1950s, signaling their low value in the eyes of the nation that bore them. There has recently been publicity about the sterilization laws in liberal democracies. Some of the laws established before World War II in Scandinavian countries and Switzerland were not repealed until the early 1970s. In one of the Swiss cases of the 1940s, and typical of many, a young woman was forced to have an abortion at two months of pregnancy and then to be sterilized because she was deemed by the investigators to be "feeble-minded, morally weak, idiotic, and promiscuous." It is discreditable that so many years after eugenics had been relegated to the scrap heap by biologists such laws were still being used to exercise prejudice. Unsurprisingly, some of the victims still alive today are now seeking compensation.

In Germany, however, events took an even more ugly turn. The Swiss law of 1928 in the canton of Vaud was fashioned into the law for Prevention of Hereditary Disease in Posterity in July 1933, when the Nazi Party came to power. A campaign for racial hygiene began with the purpose of eliminating birth defects. Doctors who were members of the Nazi Party presided over genetic health courts, and although few had much, if any, formal training in genetics, they pronounced over cases of deformity, schizophrenia, and mental retardation; men and women who were deemed unworthy to contribute their genes to the next generation were sterilized or euthanized.

The rest of the tragedy of social eugenics became known as Hitler's Final Solution and is too well known to recount here, along with its manifestation as ethnic cleansing in other places since World War II. But the earlier American quotation about "social kidneys" is telling. Throughout the 1920s and 1930s it was thought that unless all parts of the body of society were functioning healthily everyone would suffer. Safeguarding fertility and child quality were expected to improve the human condition overall, along with better housing, education, and healthcare. But any respectability that eugenics ever had died in the Holocaust, and the slower burnout of Stalinism left a legacy of determination to protect the rights of individuals. These sensibilities went hand in hand with the free-market economic policies popularly known as Thatcherism and Reaganism. We now live in a paradoxical world in which society gives a lot of freedom to people for investing in technology to make the finest babies they can afford while at the same time worrying that this could be used to win a selfish advantage in a competitive society.

Few states today dare to dictate who may or may not marry or have children. Yet the obstacles put up against infertile people stand out as a lingering prejudice. Doctors in many countries are torn between compassion for their patients and their obligation to observe legal restraints on ART and to avoid practices that would offend public taste. In South Australia, the Reproductive Technology Act of 1995 forbids "treatment if criminal charges or history of sexual or violent abuse [have been brought against would-be patients] . . . or if [the subject has] any disease or disability which would interfere in the ability or capacity to parent a child." Echoing concerns early in the twentieth century, people still ask the question, "Surely there are good reasons why God has prevented infertile people from breeding?" Yet a simple calculation, based on the numbers of infertile patients and the impact of passing any "bad genes" to the next generation, proves that there is no need to fear that overcoming infertility will ever ruin the species.

Perhaps because the People's Republic of China does not share the same guilty past, international criticism of its government's eugenic experiment has held little sway. In enacting the Eugenics and Health Protection Act in 1993, the country launched a program to "put a stop to the prevalence of abnormal births and heighten the standards of the whole population." In a burgeoning population of over 1.2 billion, the need to contain the social costs of disability are understandable, even if the methods are not. Old fears about excessive breeding of poor stock are resurfacing in China, encouraged by the one-child families that prevail in Beijing and Shanghai in contrast to the rural hinterlands, where two or even more is the norm. Under neoeugenic laws, prenuptial couples are checked for a history of heritable problems. Prenatal screening is also required so that "abnormal" fetuses can be medically terminated. Unfortunately, this has been used as an excuse to avoid the birth of girls, even though abortion for sex selection is officially forbidden.

China's eugenic legislation created a dilemma for the organizers of the International Congress of Genetics, which was held in Beijing in the summer of 1998. A wholesale boycott by many of the world's leading geneticists was narrowly avoided. Western geneticists—even ones of politically divergent opinion—condemn eugenics not only on the grounds that it is discriminatory, but also because it does not work in practice. Selective breeding can never eliminate the pool of harmful *recessive* mutations in the population. Individually, most of these mutations are rare, if considerable on aggregate, and probably no more frequent in China than elsewhere. Human beings should never be treated like cows or chickens for ethical reasons; besides, the interval between generations is too long for changes in gene frequency to show up quickly in our species. Even though the elimination of heritable diseases would be welcomed everywhere, we cannot agree on the desirable features of human biology and behavior we should select for. Had "problem parents" been prevented from having children, we might have lost some of the world's greatest geniuses, like Beethoven, Goethe, and Byron.

Few people now live under the delusion that social engineering will ever produce perfect babies. The finger of blame for society's ills has turned away from disadvantaged groups, and we are more conscious of our collective responsibility to combat the unhealthy environment that could damage our biological legacy. The products of technology are now held to be the chief culprits of poor health and reproductive failure—including "unnatural" foods, modern drugs, toxic waste dumps, radioactivity, and, most recently, the genetic engineering of food plants and animals. Environmental pessimism has mushroomed since the early 1960s when Rachel Carson's *Silent Spring* first stimulated awareness about the pernicious effects of pollution and the destruction of natural habitats. In the foreword, the biologist who helped to found the World Wildlife Fund, Julian Huxley, wrote, "Negative eugenics has become increasingly urgent with the increase in mutations due to atomic fallout, and with the increased survival of genetically defective human beings, brought about by advances in medicine, public health and social welfare." Pessimism about the future seems to be an enduring characteristic of the human condition, though anxiety is not always fully justified.

Fears for future generations after Chernobyl and other man-made disasters appear to be largely exaggerated, though there has been an increase in thyroid cancers, miscarriages, and sperm anomalies among people living close to the nuclear plant. Other areas of Europe over which the radioactive cloud passed will produce an increase in cancers and miscarriages over the following two to three decades that will be hardly detectable compared with the millions of cases that will occur "spontaneously." The reproductive system has a wonderful capacity to endure insults and still produce a seemingly unblemished baby. Even cancer patients receiving huge doses of mutagens in the form of chemotherapy and radiation have not been shown to produce more children with birth defects than usual, though they are more likely to be made sterile as a result of their treatment; nevertheless we should be vigilant. The body naturally destroys most seriously defective eggs and sperm before conception and many abnormal embryos and fetuses afterward, and these screening processes safeguard the genetic interests of the next generation.

Such processes were always necessary, even in a pristine environment, because cells can generate chromosome errors during division and produce harmful "free radicals" during normal metabolism, which can damage DNA. The role of environmental harm in producing birth defects is in addition to the huge amount of waste that occurs naturally. Only a tiny minority of the several million eggs that are made in the ovaries or the 4 trillion sperm made in the testes during a lifetime will ever make a baby. Even the destiny of the successful embryos is far from assured: more than half their number perish in the early stages of pregnancy. Although many of those eliminated are abnormal, nature has never yet come up with a foolproof screening system: a great many imperfections slip through the net even in pregnancies of the young and healthy.

One in 20 babies is born with some form of congenital abnormality, and a little under half of these are regarded by pediatricians as serious. One in 100 babies has a disorder caused by a defective gene, and one in 200 has a chromosome imbalance. Other disadvantages of our genetic legacy may not turn up until much later when diseases that reduce the enjoyment of life or shorten its length emerge—arthritis, diabetes, cancer, and so on. While the risks vary slightly from place to place and in some cases increase with parental age, it may seem surprising that natural selection has failed to come up with a more efficient screen. Why shouldn't all parents be able to count on having a baby who gets off to an excellent start in life, free of inherited disabilities and with full assurance of health and longevity?

Unfortunately, or perhaps fortunately, evolution does not operate methodically. The outcomes of natural selection are neither predictable nor perfect; the lottery of reproduction provides opportunities for both healthy and harmful genes to survive. But who are we to say that nature's mistakes have not been prices worth paying?

Celebration of Variation

In its original state in the Garden of Eden, the whole realm of nature was believed to have been "good." Only when the serpent came along and Adam and Eve ate the forbidden fruit did God's original plan go awry. In the Flood, Noah packed a pair of each species of creature into the Ark so that they could seed the world when the waters subsided. According to creationism, each species was an immutable and precious work of creation that bred true. Any departures from the original design were thought to be imperfections that had never been intended by God. Over the centuries this provided a cruel justification for maltreating disabled or disfigured people.

How different the world looks since Darwin! He confirmed the suspicions of his grandfather and others like Lyell who realized that the living world must have come into being long before Archbishop Ussher's estimate. Countless types of animals and plants had come and gone before humans arrived on the scene. Darwin's special contribution was a theory that explained why the changes had happened. He discovered that the marvelous range of plants and animals on earth and their wonderful adaptations to their environments were due to chance variations and natural selection of the fittest.

Evolution never "aimed" to make anything, least of all an ideal species or a perfect baby in any classical sense. Even the idea of "progress" is scientifically dubious: the concept of inexorable advance in complexity, beauty, and intelligence is one of the popular fables about evolution. Cultural progression has certainly occurred from wheel to space shuttle and from cuneiform tablets to computer software, but the living world never graduated to greater degrees of biological sophistication. J. B. S. Haldane once quipped that although he didn't know what the Creator was like, he was sure that he must be overly fond of beetles (half a million species). But had every type been advancing steadily, all the "simpler" models would have been left behind in earlier epochs. Darwin would have derided the idea that the advances in Victorian times that gave us iron bridges and electricity mirrored the natural world. Evolution was all about adaptation and survival, never about progression. What mattered most was the ability to adapt sufficiently well to local conditions in the environment so that a plant or animal could produce descendants. He would have scoffed, too, at the suggestion that nature could come up with a designer baby, except as an invention of the evolved human mind.

The Harvard geologist Stephen Jay Gould, with characteristic perspicacity and gusto, also attacks the myth of progressive evolution. In his book *Life's Grandeur* he points to example after example in the fossil record of species that appeared to be strong, elegant, or clever but perished because of their failure to adapt to changing circumstances. Extinction occurred not because of genetic imperfection but because the species was plain unlucky. The saber-toothed tiger and the Irish elk died out not because their teeth or horns grew too large but probably because they were ill fitted to the threats and opportunities in the environment, such as climatic change or a new predator. Evolutionary success has nothing to do with making paragons of beauty and design. Some species have evolved into far from pretty sights but are striking success stories nonetheless. Gould quotes the example of *Sacculina,* a parasite that opted for mobility on the undersides of crabs, leaving its barnacle cousins sticking to the rocks. *Sacculina* has become reduced to little more than a reproductive sac, but this has served the species well.

Even closely related breeds of animals that have become adapted by natural and artificial selection to a distinct environmental niche are exchanged at peril. Genes that are perfect in one environment can spell disaster in another. Take a tough cow living in the African veldt and swap it for a plump Jersey cow that produces ten times the volume of milk. The African cow will double its production on the damp pastures of southern England, but the Jersey, which is assumed to have superior qualities, will soon die on the veldt of diseases for which natural selection never prepared it. By the same token, only the protective environment of domestication and its high market value have saved the Belgian blue cow from extinction—this breed that nature would have eliminated because of its obstetric problems survives only because it is economically worthwhile to produce a double rump-steaked animal at the cost of delivering its calves by Caesarean section. There can never be a genome that is ideal for all circumstances.

Variation in the wild state is necessary if evolution is to occur at all, though this must be heritable if selection is to act on favorable types. Some of the evidence that Darwin found most convincing was the artificial selection by pigeon fanciers of oddball varieties—fantails, baldheads, jacobins, brunswicks, nuns, and turbits. But variation has more usually been regarded as a flaw in nature, even though it occasionally turns out to be for the evolutionary good of the species. If the Reverend Robert Burton had peered out of his study at Christ Church and seen a pure albino blackbird in his college gardens, he would have regarded it as a poor specimen because albinos are more vulnerable to predators. But if ever long, snowy winters returned to Britain, albinos might well have an advantage over normal blackbirds. Better adapted to evade predators during the winter, their numbers might gradually increase until the land became filled with white birds.

Something very like this has actually occurred over the past century in the moth population and provides a classic example of natural selection in action. Due to a mutation in a gene associated with pigmentation, a dark form of the normally light-colored peppered moth appeared in England during the industrialization of the Midlands and then in other parts of Europe. As pollution began to kill lichens, the dark form was overlooked by birds searching for a meal on bare brown tree trunks and proliferated, while the pale form was disadvantaged by its loss of camouflage. The pale form will reappear as pollution decreases and lichens return. Wherever there is a variable environment, we can expect to find a corresponding variation in animals—in shape, color, or physiology. The features this variation comprises are known to biologists as the "phenotype." While evolutionary changes have to be based on the inheritance of genes or DNA, it is the phenotype that matters and is favored by selection in nature. In short, deviations from the norm can be beneficial, although it is chance that dictates which ones are the winners and which the losers.

Healthy variations, or "polymorphisms" as biologists call them, are commonplace, and never more obvious than in the color of skin, hair, or feathers. Body surfaces are just the most visible of a string of differences that can be detected biochemically by analyzing DNA and protein. Modern humans emerged from a single stock coming out of Africa about 100,000 years ago. As our ancestors gradually spread to all the continents except Antarctica, genetic differences grew like the branches of a tree, and linguistic divergences emerged in parallel. These branches formed relatively recently in evolutionary terms, and our species remains fairly homogeneous, contrary to prejudices that used to consign races to separate subspecies or even distinct species. The Harvard biologist Richard Lewontin estimated that if only the Kikuyu tribe in Kenya survived the extinction of humankind, 85% of all human genetic variation would still remain. Differences between ethnic groups are slight when set against the variation that exists across the species. In light of this fact, any objections to using donor eggs or sperm for people of different races are biologically as absurd as the old marriage laws under apartheid in South Africa. Scientific progress has provided more grounds than ever before to deride the use of the term "race," which implies more differences than are actually warranted by the evidence of human affinity. The Victorians have often been criticized for being class and race conscious, but few people lambasted old prejudices more incisively than one of the great thinkers of the age, the economist John Stuart Mill: "Of all the vulgar modes of escaping from the consideration of the effect of social and moral influences upon the human mind, the most vulgar is that of attributing the diversities of conduct and character to inherent natural differences." How far bigotry and suspicion about foreigners have disappeared is debatable, though there is no doubt about the progressive influence of international travel, immigration, telecommunication, and education for human understanding. The cost of these opportunities is a potential loss of human diversity, though more cultural than biological. A more uniform global culture is developing, and people are sharing more of the same aspirations, from Levi's jeans to satellite TV—and perhaps a standard baby.

The genetic melting pot is being stirred as never before, and this has been beneficial for warding off the risks to health and fertility of inbreeding. In the Middle Ages, people rarely traveled farther than ten miles from home, and the choice of marriage partner was correspondingly restricted. Marrying of close relatives was forbidden, though nobility and royalty sometimes chose to make matches in related families to keep the family silver together. Darwin, a member of the minor gentry, married his first cousin, Emma Wedgwood, which was permissible under civil and canon law. But after he had finished experiments to test self-fertilization in plants and found that their fruitfulness was diminished, he became more concerned about his own marriage and bloodstock.

We generally prefer inbred or pedigree animals to those that are outbred because their characteristics are pleasing and more predictable. Even so, to be surer of a long-lived pet and lower vet bills, it is safer to go for a mongrel. Likewise, commercial pressures to produce ever higher yields of crops are leading to greater uniformity in staples around the world, and the loss of wild-type genes from the pool is worrying as these may include genes that will resist fungi and other pests in the future. Inbreeding can leave a population vulnerable, especially if it is a tiny, isolated one. For example, the Glanville fritillary is a pretty butterfly found only in England at the Undercliff on the Isle of Wight. Studies of butterflies in Finland have shown that as populations shrink, they become more uniform and vulnerable to sudden extinction as they are less able to adapt to slight changes in climate, food plants, or predators. Whereas inbreeding can be the kiss of death, genetic diversity is a sign of population health and security. As humans increasingly encroach on natural habitats, forcing animals into ever smaller breeding nuclei, genetic refreshment becomes important to ensure these species' survival. We see an example of this in conservation, where biologists ship animals belonging to an endangered species or their frozen semen between zoos or to small, isolated colonies in the wild.

Genetic variety is generated whenever an egg and a sperm from unrelated individuals meet to make an embryo and a potential new person. Each gamete cell carries one set of 23 chromosomes with genes strung out along their length like colored beads. We inherit one set from our father and one from our mother, making 46 chromosomes in every cell in our body. The gametes originally had 46 chromosomes too, but they underwent a unique process of genetic exchange and division—called meiosis—during their formation. This process starts before birth in eggs but in sperm waits until puberty. But in both cases, meiosis is necessary if they are to fertilize. The pairs of chromosomes link up, like perfectly matched dancing partners, and exchange several segments with each other (though in the testes the male Y chromosome remains rather aloof from its partner, the female X). This process of genetic recombination is followed by the cells splitting to reduce the chromosome number to 23. It guarantees that every egg and sperm is genetically unique—unlike the somatic cells of the parent's body, which are normally uniform. If we could see the "colors" of genes along the chromosomes every time a new gamete is made, they would always look different. The exchange of beads produces plenty of potential pheno-types in the next generation for natural selection to play with.

If evolutionary selection has done its work and honed the genome to make the species well adapted to its envi-ronment, most of the genes are, ipso facto, beneficial. As long as one member of each pair of genes is favorable and has a dominant effect, it matters little to the well-being of an individual whether the other one is "good" or "bad." The subordinate, or "recessive," gene becomes significant only if it forms a pair at fertilization with another of the same kind and the usual, or so-called wild-type, gene is absent. Occasionally, however, a gene may be altered by mutation to make a stronger, or "dominant," form, and this could be harmful. As its label suggests, a dominant mutation is expressed regardless of whether its partner is the healthier for the current environment.

In our example of the albino blackbird, the mutated gene for albinism (there is, of course, no "albino gene") is recessive because it fails to have any effect when a normal gene is present for the making of pigmentation. The indi-vidual possessing one copy of each gene is called a "carrier" because the mutant form does not reveal itself in the phe-notype, though it might show in the next generation if per-chance two carriers mate to produce a pure albino. This principle holds true for other complementary pairs of genes, too, whether one has relatively innocuous effects like albinism or causes disease.

Cystic fibrosis is the most common inherited disease in people from northeastern Europe. Nearly one in 20 of the population—both males and females—carries a copy of a harmful recessive mutation for a gene affecting chloride transport in cells, which, if it is expressed, can have dread-ful consequences for breathing and digestion. They are likely to be unaware of it unless they have been genetically screened or have family members who have suffered. The likelihood that carriers will marry, assuming random pairing, is 1 in 20 times 20, or 1 in 400. Now, what are the chances that their offspring will have the disease? Remember that meiosis splits the chromosomes to make an equal number of eggs or sperm with either the normal or the mutant gene. If these gametes are mixed, there will be a 25% chance that the couple will produce a completely healthy child with two normal genes, a 50% chance of a carrier who is otherwise healthy like themselves, and a 25% risk of a child who inherits both mutations. This 1:2:1 ratio was discovered by the Moravian abbot, Gregor Mendel, from experiments on peas in his monastery garden. Mendel grasped a principle that is so general that it is used by genetic counselors to predict the risks for many inherited diseases in humans. In theory, the frequency of cystic fibrosis should be 1 in 4 times 400, which is not much different from what has been found in the population at large. Increased awareness coupled with prenatal diagnosis can reduce the incidence of affected babies, though without much affecting the frequency of recessive genes in the population.

One defective copy of a gene can sometimes be enough to cause a disease if it is dominant, regardless of what its opposite number is like. Huntington's disease requires only one gene to be affected for the seeds of the devastating condition to be sown, though the symptoms do not appear until midlife. Like cystic fibrosis, it became possible to test prenatally for Huntington's disease in families at risk once the gene was identified. Some harmful mutations, although recessive, behave as if they are dominant when they occur in a boy. This is because they are carried on the X chromosome, and males carry only one copy. Males are much more vulnerable to a disease emanating from a fault on this chromosome, and these diseases, including hemophilia and Duchenne's muscular dystrophy, are called "sex-linked."

Such examples are so clear-cut that the complexities of the genetics of many inherited diseases are sometimes overlooked. Some potentially harmful genes are described as "nonpenetrating" because they do not affect every individual carrying them. These contingent diseases require a coincidental factor to be triggered. This could be an infection or a drug or something in the diet, just as smoking puts those inheriting a certain kind of gene at much greater risk of lung cancer. The prospects of long and healthy life cannot be predicted from an individual's genome without knowledge of the environment in which the person is living. As the circumstances of modern life are constantly changing, it is impossible to predict what set of genes will be in the best interests of our grandchildren living in an environment different from our current one. Indeed, we still do not know enough about our biology to say what genes are best for us now.

Some inherited diseases are due to genes that once served useful purposes and were favored by natural selection because the price was worth paying. They force us to revise what we mean by a "bad" gene. Of course, for the mutation to spread through the generations the benefit need not be good for everyone as long as there is a net advantage for the majority of carriers. Most famously, the sickle cell mutation is present in 20% of the population in parts of Africa and in 5% of African Americans. It is due to a recessive mutation in the gene for a protein chain in the blood pigment hemoglobin. When two copies of the mutation are present, sickled blood cells start to block the fine blood vessels, causing debilitating effects and leading to the full-blown disease—which eventually proves fatal. But carriers of the disease, with only one copy of the mutation, have the advantage of resistance to malaria without its serious consequences. This enhanced the survival of these people, and therefore of their genes, in the past. Genetic disease is sometimes the biological price of evolutionary opportunity, but this knowledge is of no comfort to parents and their children burdened by disability. This was the grim truth that prompted Charles Darwin to write to his friend Joseph Hooker, perhaps thinking of the loss of his little Annie: "What a book a devil's chaplain might write on the clumsy, wasteful, blundering, low, and horribly cruel works of nature."

Another condition, which is not so serious but affects thousands of women around the world, is polycystic ovary syndrome. Although the gene—or genes—responsible for this condition has not been identified yet, this syndrome is known to be linked with insulin resistance and a tendency to obesity and diabetes. It is not stretching the bounds of credibility to suggest that the "thrifty" gene responsible helped our ancestors to survive famine by storing energy and suppressing fertility.

The list of mutations that are supposed to have been "virtuous" in the past and are now harmful is long and controversial. Far more genes have alternative forms or effects, which are called polymorphisms, that are innocent and of which we are often unaware. Take a random sample of people on the street and invite them to taste a test paper impregnated with phenyl thiocarbamate. Only a few will be able to detect the chemical, revealing that they possess a particular form of a gene. There is no advantage in having this gene today, but at one time it may have stood people in good stead when there was a risk of eating foods that cause a goiter. Our genomes are full of such baggage from the past, but there is no more biological urgency to tidy it all up than to clear out a dusty trunk Grandpa left in his attic. The old clothes and newspapers he stashed away may be of little relevance to us today, but if they are not taking up needed space, why not just leave them there and get on with our own lives? Perhaps they may even turn out to be valuable in generations to come.

Even if we could "tidy" our genomes and rid them of the clutter of DNA that does not function as genes, it might do more harm than good. Junk DNA can harmlessly absorb a lot of damage from chemicals and radiation that otherwise disrupts gene action. Replacing "useless" gene variants with others we think are better might be risky. Besides, advances in healthcare, diet, and medical treatment are constantly redefining what is beneficial. Nearsightedness and astigmatism are serious disadvantages for a hunter-gatherer, but we can treat them cheaply and effectively with spectacles. Phenylketonuria, or PKU, is a serious inborn metabolic disease: an inability to convert the amino acid phenylalanine, which causes profound mental retardation. But the screening of newborn babies can identify affected individuals soon enough to prevent the onset of the disease through the prescription of a restricted diet throughout childhood.

Successful treatment of any inherited disease that allows affected individuals to reach the age of reproduction inevitably increases the frequency of the gene in the next generation. This has been a cause of concern since Galton's day. Some individuals choose to be screened for inherited disease and forego fertility if necessary to avoid having an affected child, but there can be no justification, either morally or scientifically, for returning to a social policy of reproductive exclusion. There are many types of genetic fault in human genomes, but most of them are very rare, and some, like those mentioned, are no longer so threatening in the new environment we have created for ourselves. There are, of course, social and financial costs for treating people with medical conditions, but these are small prices to pay for the freedom to choose what is important to us.

Stone Age Legacy

Charles Darwin is remembered less for his concerns about the disastrous experiments of nature than for what he had to say about another hard fact—"the struggle for existence." Animals and plants produce more offspring than they strictly need to replace themselves, but unless there is selection there can be no evolution. What matters in the evolutionary stakes is ensuring descent: the animals that were most successful at this were the ones Darwin called the "fittest." This is true whether the animal is a prodigal r-type breeder, like a cod, or a K-type, like a human being. Hence there have often been attempts to apply evolutionary theory to human behavior, and this has proved most provocative.

The theory never found much room for altruism between unrelated individuals, although hard-hearted patterns that proved successful in nature should never bind us to a model of human society and behavior. Animals groom each other, but such tokens of cooperation are now regarded by biologists as services provided for something in return, and usually only for relatives. The optimistic view of animal self-sacrifice has gone, too, like the kamikaze pilots of World War II. The old story about lemmings throwing themselves into rivers to drown so that the population was kept in check in time of food shortage does not hold up in either field observation or theory. There is a story that filmmakers had to sweep lemmings into a river to obtain the shots they wanted, demonstrating our fondness for theories even after they have ceased to hold water. It does not take a genius to work out that an individual who acquired a genetic tendency to commit suicide in the interest of the group would be at a selective disadvantage. Those who were left would get on with the job of putting their genes into the next generation with their usual self-interested enthusiasm. No matter how seemingly beneficial for the species, selfishness seems to prevail. This is all too often the case in our own species, and we are at our most generous when it comes to our own offspring and to people who share our genes.

In nature, tough competition and genetic jealousy often reign. When a male lion takes over a pride, it kills or evicts the litter of its new mate to avoid having to invest energy in the young from another father's genes. The strategy pays double dividends because their females return more quickly to sexual heat so that the new male can impregnate them sooner with genes of his own. In evolutionary parlance, the thwarting of the competition is called "reproductive spite"; its sociobiological counterpart is "kin selection." These two phenomena form the basis of much of what follows, but it is important to bear in mind that there is no full agreement among biologists on how much relevance they have for anthropology today.

Reproduction is a proxy for immortality. The genes we inherit—some of them going back to early epochs on earth—are passed on to our descendants in turn. The body has been likened to a vessel for copying our genes and carrying them on to the safety of the next generation before the body founders on the rocks of old age, if not earlier. This analogy has been popularized by the Oxford biologist Richard Dawkins in his best-selling book *The Selfish Gene*. It brings to mind an aphorism attributed to Samuel Butler—that a hen is an egg's way of making another egg—and biologists too can slip unconsciously into teleological ways of thinking. We may find this counterintuitive way of looking at nature repellent, but the reasoning is nonetheless very persuasive. The world of nature we so idealize is not progressing toward harmony and perfection as we prefer to hope. Evolution is blind to our happiness and security. Natural selection blunders on, coming up gains in health, "brains," and physique only when they are in the interest of passing on the genes. These genes give not a whit for our cares and, even if they cause us heart disease or cancer later in life, go merrily on, so long as they can perpetuate themselves. Like viruses, they use their hosts for reproduction, although in this activity the survival vehicles are fully cooperative.

The idea of genes selfishly replicating themselves has been extended by analogy to cultural evolution, for knowledge, art, traditions, and human values transmit themselves from generation to generation in the same fashion. The term "meme," or memorable unit, was another of Dawkins' clever inventions: this token of inheritance is the cultural equivalent of the gene. A meme can include almost anything that is passed on from one generation to the next, including ideas, tunes, jokes, even the concept of the selfish gene, for scientific theories are memes, and scientists and their conferences the agents for spreading them around the globe.

Social norms of human reproductive behavior have been transmitted as memes, too, and have helped families to adapt and make the best of their circumstances, even if they are no longer appropriate today. Whether transmission of this behavior is by memes or genes matters less in understanding what follows than the fact that self-interest in reproduction is an ongoing characteristic. While some of the dreams about making an ideal child reflect the fashions and values of contemporary society, other concerns are deep-seated and seem to well up from our past.

We need go no farther than our backyard to find natural examples of how animals and birds manipulate their chances of becoming successful parents. Studies of great tits and tawny owls have shown that they can adjust the numbers of eggs and the sex ratio of the brood according to circumstances. We do not always know how they manage to make these changes, but it is clearly a strategy they play to give their offspring the best chance of winning a mate and producing descendents to carry forward the selfish genes. Striking and sometimes horrifying human parallels can be drawn, and evolutionary psychologists suspect that similar motives are afoot.

Like a lion that rejects its rival's offspring, an Ache hunter-gatherer tribesman in South America will openly refuse to raise any child a new wife brings into the marriage. An infant's chances of dying after the departure of the biological father are 15 times higher than they would have been otherwise. These cruel attitudes are not confined to so-called primitive and backward tribes; police investigators in our society are trained to suspect stepparents in cases of child abuse and homicide. For example, in the notorious case of an English head teacher, who was convicted of murdering his 13-year-old foster daughter in 1998, he went on a shopping spree after the crime with his natural daughter as if nothing had happened. A Canadian study showed that a stepparent was 65 times more likely than a natural one to be responsible for infanticide, and this tendency has been reported in other countries.

Infanticide is the extreme tip of an iceberg of reproductive jealousy that more often finds expression in neglect and abandonment and sometimes in the abortion of an unborn child of a woman's previous partner. Over the years, with advances in medical care and changes in family structure, the balance of responsibility for offenses has shifted from women to men. When maternal mortality was high, the incomer was usually a stepmother, and in eighteenth-century Ostfriesland, a widower's child was less likely to survive if the father remarried. This recalls the wicked stepmothers of the fairy tales of the Brothers Grimm, though they generally made happier endings. The Reverend Burton expressed the problem with typical bluntness: "A stepmother often vexeth a whole family." Their general reputation has aggravated the difficulties of devoted stepparenting, which sociobiological theory is at a loss to explain. One of the most popular contemporary English writers, Joanna Trollope, describes her experiences as a stepmother as "like an unexploded bomb in the briefcase of the man you marry." Today, however, the new parent is likely to be a man since children more often stay with their mothers after separation and divorce, so there are more stories about wicked stepfathers. Within a decade from now, there will be more stepfamilies than birth families in the United States and Britain, but thankfully, there have always been far more examples of warm and tender step relationships than cruel ones. Evolution helps to explain the darker side of human nature but should never be used to condone it.

How far this Stone Age legacy affects attitudes to fertility treatment using donated eggs or sperm is an interesting question. Many ethnic groups and some world religions, including Islam and Judaism, avoid the problem by forbidding such treatment, and even in the West gamete donation services are provided with discretion and often in an atmosphere of secrecy. There is, however, an emotional solidarity between partners who agree to introduce a third party into their union that may bode well for the family.

Gametes have usually been donated anonymously by unrelated people, although arrangements with family members, friends, or acquaintances are sometimes preferred. Despite a huge expansion in the past two decades, couples find it embarrassing to talk about their need for this service. Most people who have been conceived using these methods are still not told about their origin by their parents, although the Human Fertilisation and Embryology Authority in the United Kingdom is bound to inform them when they reach adulthood if they request information. However, the name of the donor remains confidential— unless the law is changed—even though many people want to know the identity of their "lost" genetic parent.

British egg donors provide gametes gratis and sperm donors receive only a small fee, if any. Most are interested in whether their gift was successful and how many times. A legal limit of ten children after donor insemination has been set in the United Kingdom to minimize the risk of them marrying and unknowingly committing incest. In the United States, where a fully commercial system operates, would-be donors advertise their services over the Internet and through specialist agencies. They often make highly personal statements, like the man who described himself as "very sensitive open, warm-hearted, sweet, generous, nurturing, intelligent" and whose personal profile ran to four more pages! In Britain, only a bare minimum of details are provided. There are basic health safeguards: potential candidates are screened for transmissible disease, semen donations are quarantined for six months, and family histories are monitored for inherited diseases. Prospective parents can choose from a menu of ethnicity, height, body build, complexion, eye and hair color and facial appearance, but there is no information about scholastic performance and IQ, religion, criminal record, sports interests, and hobbies and the host of personal details that are available to their American counterparts. Americans are discerning shoppers in the reproductive marketplace, and the growing tendency for donors to be willing to be contacted by the couples they help or, in due course, by the resultant children must surely be a welcome alternative to secrecy in the family.

In addition to a strong preference for our own genes, other biases about our children have been passed down from previous generations. The firstborn is traditionally a favored child—especially the "sons of your youth," as the psalmist put it. In Judaism, the one "that opens the womb" was held to be holy, and if that child was a son, he inherited a double portion of the estate as well as the father's title and prestige. The threat of the last plague at the original Passover when the Hebrews wanted to leave Pharaoh's Egypt was all the more grave because firstborns were most at risk of succumbing to dread disease—perhaps because their responsibilities took them to places of infection. Not only did it make biological sense to invest most in a child born while the mother was in her obstetric prime, but also the child who first passed safely through the hazardous years of infancy would seem the most likely to succeed the father and was therefore all the more precious. There is no longer any practical reason for firstborn partiality, although the evidence suggests that it still persists because eldest children achieve higher marks at school and enjoy better health than their siblings.

The most widespread and pernicious bias has been a preference for boys over girls, although there have been a few examples of the reverse. According to the Trivers and Willard hypothesis, which is named after the authors of a famous 1973 paper, people invest more in one sex or the other depending on which pays the better dividend in family esteem, wealth, and numbers of grandchildren. For a wealthy family, it makes sense to prefer sons because male fertility is unconstrained by the biological limitations of pregnancy; where the culture permitted it the sons could have more wives and expand the family's property and influence. This preference for male offspring would be particularly advantageous where male mortality was high, as the scarcer sex would have a larger choice of marriage partners.

Such a bias in favor of male offspring remains strong in India and Southeast Asia, where from time immemorial the mandarins regarded girls as "not worth so much as boys." The dowry system is often blamed, but ailing parents also look to their sons for assistance and to take care of their funeral arrangements. In the West it has tended to be daughters who provide these services. But even here, male preferences still exist in mild forms. A recent study showed that in Philadelphia—and probably it is generally true—better-off parents with a son breastfed longer and so increased the interval to the next pregnancy. Contrariwise, Hungarian gypsies give preference to their daughters. For these poor families, a daughter is a better bet since females are more likely to marry native Hungarians above their own social station, and a higher status son-in-law is economically beneficial to the whole family.

The motives for reproductive investment probably go back at least as far as the Stone Age, from which some tribes in Irian Jaya, New Guinea, have hardly emerged. Living in a harsh highland environment, the Eipo select the babies they believe have the best chances of surviving. Infanticide was common until the arrival of missionaries 30 years ago. After delivering her baby in the bush, a mother leaves its umbilical cord attached until the afterbirth emerges some 20 minutes later. During this time she observes the child for life signs and examines it to decide its fate: if it seems healthy, then good; if it's a male, better still. Clubfoot is a serious disadvantage in the mountainous terrain, and any child with such a disability is likely to be killed; a baby conceived extramaritally is also quickly despatched and buried. Healthy sons are never killed, but if the mother is over 30 years old and has only daughters she may eliminate a daughter and hope for a son next time around.

Although abhorrent to our eyes, child killing is not always the wanton and indiscriminate act it is often portrayed. The practices of Eipo tribeswomen are more understandable through sociobiological theory and bear comparison with abortion practices in other parts of the world. The baby is killed while the mother still considers it to be part of her own body and before maternal bonding has occurred. A son is a bonus in a society where 25% of men die violently, polygamy is common, and the number of females sets a limit on the birth rate in the community. Women invest prudently in what they have been taught is best for their family, and in the precarious existence of this tribe their attitude helps to keep the population stable.

The Eipo people, like any group in our species, are intelligent and sensitive, and their motives and choices, in terms of evolutionary psychology, are understandable, even if we recoil from their actions. Investment in a child who can cope in a tough environment and eventually bring the joy of grandchildren into the family is more understandable than some of the choices that are being made, albeit before birth, in Western consumerist societies. We do, however, have something in common with these people in the way we can hold sway over the characteristics of our children. The most natural example of this is that our choice of sexual partner decides half the genes of our future children, although this may never have been at the front of our minds during courtship. Biology indicates that sizing up a potential mate is not simply a matter of aesthetics.

Cupid's Arrow

There is an old assumption that a well-made marriage makes the finest issue, just as the best ingredients are used to bake an excellent wedding cake. But there has never been much agreement about how such a match should be made or its quality measured or even who should make the choice. The final decision is important to the rest of the family since it determines the composition of the gene stock that will be passed on to the next generation and who will have the property rights. With these interests at stake, it is not surprising that parents have often coerced their sons and daughters into arranged marriages and applied penalties for resisting their wishes. In *A Midsummer Night's Dream*, Theseus advised fair Hermia to reconsider her decision to reject the suitor her father had chosen for her or she might end up "withering on the virgin thorn."

But setting such arrangements aside, where we are free to choose, the manner of falling for someone varies enormously. No one has come up with a fully satisfactory theory to explain the essence of love and physical attraction. Most of us enjoy the mystery, like the gambling Cavalier in the court of Charles I, Sir John Suckling:

> *But love is such a mystery,*
> *I cannot find it out:*
> *For when I think I'm best resolv'd,*
> *I then am most in doubt.*

He would have been even more bewildered today because the choice of partners has never been so great. Travel, high school or college, and the workplace are some of the commoner circumstances of first encounter, but for those who are less fortunate or just too bashful to speak up there are introduction services provided by a plethora of agencies. Most people would probably deny that their criteria for choosing their partner had anything to do with the prospects of successful reproduction; instead they would cite a pleasing appearance, talent, youthfulness, good humor, or financial security. The drive to make heirs may seem less urgent these days, but love, with its underlying sexual chemistry, has evolved to ignite the reproductive gunpowder. In a sense, perhaps we have always been trying to make "designer babies" by tipping the odds toward a favorable reproductive outcome through choosing the best mate available.

Choosing the right partner is more important for females than it is for males, at least in nature. A woman has fewer breeding opportunities than a man, who can potentially breed with many mates. That is why traditionally males do the courting and females the choosing. The same reason explains why males of many species have a more gaudy appearance than females; the more competition a male faces, the more striking his appearance is likely to be. Among polygamous species of birds the male is almost always the brighter of the pair. On the other hand, in species that mate for life, like swans and albatrosses, the sexes are much more alike. Darwin called this showiness among males "sexual selection" and must often have been reminded of it as he heard the peacocks calling in his garden at Down House.

A life of polygamy may sound attractive to some males, but the process of sexual selection brings stresses and dangers with it. A striking appearance may attract mates, but it also attracts predators, and an extravagance like a peacock's tail has economic costs as feathers are broken and have to be replaced and parasites have to be cleared. Studies of barn swallows demonstrate just how cautious a polygamous male must be. If his tail streamers are artificially lengthened by adding extra feathers, his aerial display to females is more successful and he raises more chicks. By extending them just a little further he runs into aerodynamic problems and is less adept at capturing prey and evading hawks. There is a trade-off between safety and sexiness.

A century after Darwin, Bill Hamilton of Oxford University pointed out that bright-colored birds are showing off to choosy females that they possess good genes. A bird displaying immaculate plumage free of parasites is demonstrating his genetic merits to a female concerned to enhance the quality of her own investment in reproduction. Appearances are not everything to females though, and they will often choose an older male if he has proved himself to be powerful and successful. A stag fighting and bellowing during the rut, for example, shows the quality of his genes and his ability to defend a harem. According to a Brazilian saying, "Good looks don't put anything on the table"; this is as true in the wild as in human society.

It may seem a long stretch from animals to humans, but our species too is competitive and was never strictly monogamous. What is more, testosterone is the driving force of male-type behavior. When a young man gets into fights or drives recklessly, he could be sending signals as surely as the bellowing stag, declaring to females that he's "got what it takes."

The stakes are so high that females are rarely passive onlookers. Flirting has a biological purpose for by encouraging more than one suitor at a time, the genetic worth of the eventual winner is likely to be greater. This is logical in sociobiological theory, and there is plenty of evidence of ménages à trois in species as different as garden birds and gibbons. Formerly presumed to be monogamous, these species, along with a large number of others that are conspicuously polygamous, are now known to court third-party involvement. Sperm competition is an insurance policy against ending up with a poor male, although most males solicitously guard their "wives" against being cuckolded. Research has shown that extra-pair copulation occurs more frequently in humans during the woman's fertile phase, whereas sexual activity within a pair is more evenly spread throughout the cycle. There may be more to lovers' games than is generally thought.

Odor is one of the means of exchanging signals between members of the opposite sex. Any odor that is transmitted by one person to affect the behavior or physiology of another is called a pheromone. In some respects, it is an external equivalent of a hormone, often acting in the same subliminal way. Smell is our most archaic sense, and the region of the human brain where it originates was one of the earliest to evolve. There is tantalizing evidence that the olfactory sense is tied up with another system in the body with a long genealogical record—immunity—and together they can enhance reproductive success.

The key is a family of genes called the major histocompatibility complex (MHC). It is well named, since the variety of genes in the MHC is so complicated that we nearly all have unique combinations. This is why immunosuppressant drugs must be used to prevent the organs of donors being rejected after transplantation. The MHC did not, of course, evolve to thwart the work of transplant surgeons; it acts as a sophisticated warning system on patrolling white blood cells as the MHC is able to detect invading pathogens. The variety of MHC types partly explains why some people are more resistant than others to infection and certain diseases.

It would be best for parents to make babies with different MHC types so that at least some of them would be tough enough to withstand the onslaught from the microbial world. Better still, they should make babies with the MHC type that is most effective at warding off the particular diseases they themselves are exposed to. Mice seem to find mates of a different MHC type more pleasing, probably because they smell differently. What is more remarkable, unions between eggs and sperm at fertilization have been found to be more selective than had been assumed. Mouse pups born to parents that had been infected with the hepatitis virus inherited combinations of genes for MHC that would help them fight the disease, which implied that eggs were being choosy about their partners, or vice versa. Should this appealing story force us to wonder whether in the sperm injection technique (ICSI) embryologists are choosing the best sperm for human fertilization? Perhaps not, but then we understand so little about the processes involved that the verdict should remain open.

Research is revealing many subtleties about human sexual attraction and conception. A link has been established between pheromones in men's body odor and women's sexual preference, especially around the time of ovulation. In one experiment, American college women sniffed and rated the attractiveness of T-shirts worn over two nights by men (who avoided eating garlic or using deodorants). At the ovulation time of their cycle, fertile women preferred the scent of the men with the more symmetrical appearance, symmetry being gauged by comparative measurements of ears, hands, and limbs; women who were taking the pill and not ovulating had no such preferences.

"Fluctuating asymmetry" is an index of developmental stress. It occurs when the genetic blueprint laid down in an embryo is imperfectly translated into the shape of a fetus, probably because the pregnancy was disturbed or exposed to something harmful. No creature is perfectly symmetrical, and some organs, like the heart, obviously have to be otherwise. Nevertheless, external symmetry in animals seems to be beneficial and is associated with longer life, fewer illnesses, and higher fertility. Among humans, symmetrical men are said to be more attractive to women and have more partners and children, and their sexual partners are claimed to have more orgasms.

Francis Galton discovered the appeal of symmetry for humans by accident. He was blending photographic portraits of people to find out whether there was a distinctive criminal face. Although the experiment suggested that there wasn't such a thing, he made the observation that "all composites are much better looking than their components, because the averaged portrait of many persons is free of the irregularities that variously blemish the looks of each of them." This phenomenon is now called "koinophilia" and has been confirmed in both men and women using computer graphics to construct the most pleasing appearance. Perhaps it explains the saying that attractive faces are more difficult to remember. But it should be borne in mind that the correlations are very weak. Symmetry is obviously not one of the most important things about a person, though it is nonetheless interesting to evolutionary psychologists.

While females are attracted to odor, males are more responsive to external appearance. Perhaps this is why women are shy about declaring their age and go to such lengths to preserve their looks. In Oscar Wilde's *The Importance of Being Earnest*, Lady Bracknell famously maintained that "London society is full of women of the very highest birth who have, of their own free choice, remained thirty-five for years." Whether Wilde realized it or not, this observation accords with biological reality, because a woman's fertility falls markedly after 35, and the risks of having a miscarriage or producing an abnormal baby increase. A woman's age is a gauge to her fertility and, sad to admit, her attractiveness to younger males. The reverse discrimination does not apply as male fertility is maintained until much later in life, and it is more usual for women to marry older partners.

A woman's face is normally the first thing an interested man looks at closely. The cosmetic industry makes a fortune as women make clever attempts to hide wrinkles and crows' feet, but there are other telling signs that are harder to conceal. The nose and ears gradually enlarge with age, while the eyes shrink under sagging flesh. Skin color darkens, and the lips, which serve a woman like a bird's plumage to advertise that she is not anemic and that she's therefore good for pregnancy, lose their fullness. The key indicators of youth are known as the neotenous features, and models featured in glamor magazines conform to the ideal. Surveys of five populations, ranging from Ache tribesmen to Americans, confirmed that men are attracted by a high index of neoteny.

In prehistoric times cavemen painted as many breasts and buttocks of women as they did faces, and these remain important foci of male interest, even though leanness is now more fashionable. The waist-to-hip ratio is yet another rough indicator of a woman's age. A ratio of about 0.9 in girls and boys before puberty falls under the influence of female hormones at puberty to about 0.75 by age 20 as girls put on fat around their buttocks and hips. By middle age, when fecundity is declining, women begin to deposit more fat around the waist, and the waist-to-hip ratio climbs back to 0.9. When American men were asked to select their preferred female silhouette, most chose the one with the greatest difference between waist and hips. But biological preferences are tempered by social factors. Doug Jones, an American anthropologist, has found that the physical features males find attractive in females vary considerably according to country and culture. So while men in America find large breasts attractive, Brazilians tend to focus more on buttocks and thighs. Women go for a more complex mixture in men, including status, personality, future earning potential, and, to a lesser extent, looks.

These predictions are no less controversial in their own way than those that were originally claimed for evolutionary theory more than a century ago. We cannot squeeze all we know about the human body and behavior through the mill of natural selection theory and find an adaptationist explanation for everything. But there is much to commend the cautious use of theories for reproductive behavior and sexual preferences. Above all, in nature the variations are just as interesting and important as the standard forms we take for granted, and it is through constant genetic variation that species have been able to adapt and survive from generation to generation. The temptation with the advent of new technology is to avoid what we presume to call defects and circumscribe human variation by attempting to "select" attributes of a child-to-be. The possibility and morality of such a project are key elements in the debates over producing "designer babies."

Last, we should remember that although evolution accounts for our natural mode of reproduction and our psychic drive to produce offspring, it otherwise holds little sway over the future of the species. The human brain—the pinnacle of evolutionary achievement—has now replaced the brute forces of natural selection and sexual selection as determinants of change. Our capacity to reason can override the reproductive choices we used to make, and our genius at science and invention is opening the door to fresh opportunities. Less constrained by biology than in any previous era, technology now becomes the key to determining our fertility, the type of children we create, and hence the kind of society we are making.

4

Playing God

Getting the Picture

Until recently, there was always plenty of speculation when a baby was on its way. Would it be large or small, dark or fair? Would it inherit red hair from its mother's side or the noble nose from its father's? Would it be a he or a she? Little sweaters were knitted in yellow wool to play safe. There was something cooking inside Mother that stimulated imagination and conversation among the whole family, but although everyone thought they knew what the ingredients were, nobody was sure how the cake would turn out.

This was the way generations of people regarded a pregnancy. Little had changed when my own children were born about 20 years ago. Prenatal development was still a mystery. The closest a doctor or midwife could get to examining a baby in the womb was at the other end of a Pinard's obstetric stethoscope. Once the mother had recovered from the shock of having the cold metal pressed against her belly, she could be reassured that there was life inside. Medical attention was firmly focused on the mother— checking her weight and blood pressure, analyzing the chemistry of her blood and urine.

A woman may have more health checks today than ever before, but in physical terms, of course, a pregnancy feels the same as it has always done. The big difference is that the womb is no longer a secret place. Ultrasound scanning can monitor the fetus without subjecting it to risk or invading the mother's body. Pregnancy screening has become so familiar that we rarely stop to think about all its implications or pause to wonder about the directions in which technology will leap forward in the future to reveal even more. We can, however, be sure that the baby will be more predictable in terms of its physical development, health, and prospects of long life.

Tracy is one of over 4,000 women who will deliver in our hospital this year. Her second scan at 20 weeks of pregnancy was typical. As she lay on the bed, gel was smeared on her swollen tummy to allow the scanner to glide easily in the hand of the technician. Her placenta immediately came into view on the monitor as a shady disk. Having checked that it was normal, the ultrasonographer switched her attention to the baby and gave it perhaps the most thorough internal examination it would ever have. To her trained eye, its ghostly image revealed a great deal of detail. Having checked the heart, intestines, liver, diaphragm, and kidneys, she paused to watch the bladder filling and emptying properly. All the major bones were visible, and the skull was measured to check that the baby's growth was on target. The spinal cord and brain were examined next for abnormal openings or swellings that might indicate spina bifida, hydrocephalus, or other neural tube abnormalities. Nothing is scrutinized so closely as the nervous system because we consider the quality of our mental life to be of paramount importance.

After examining the brain of Tracy's baby, the ultrasonographer switched her attention to the neck of the fetus before directing the scanner at its nether regions. This may seem a surprising area to look for trouble, but fluid accumulating in "nuchal folds" can point to Down's syndrome and a number of other abnormalities. Last, the baby's sex organs came into view, and Tracy's teenage daughter, who had joined the viewers at her bedside, was delighted to discover that she was going to have a sister. The baby was normal and fully formed, and the sonographer gave a quiet sigh. Now all that remained to be done was to take the first picture of the baby in the womb. Although not perhaps the most complimentary likeness the little girl would ever have, it would help her relatives bond with her and prepare for her addition to the family.

Such a picture serves as a memento of what is usually a happy visit. Pregnancy is a natural process, and scanning reassures everyone that the baby is growing and forming well and lets the parents into the secret of its sex. But behind the celebratory atmosphere, there is a serious intent. Just as conception never promises a pregnancy, so pregnancy never guarantees a baby, let alone a perfect one. Parents still cross their fingers and pray that everything will be all right. Most people approach scanning with little anxiety: they anticipate no problems and usually return home reassured. But those who are in an older and riskier age bracket are less likely to feel at ease during these sessions because they know that such tests could bring with them hard choices. Prenatal medical technology is better at predicting defects than curing them. If screening indicates a serious fetal defect that is then confirmed by prenatal diagnosis, they must decide whether or not to terminate the pregnancy. The moral dilemma is whether the better view of the fetus that technology affords us will lead to this decision being taken with greater or lesser respect for its humanity. What level of imperfection warrants a decision to terminate?

Maternal blood samples are used in conjunction with scanning to determine the risks of the fetus having a neural tube defect or chromosome disorder, such as Down's syndrome. The levels of certain proteins and hormones in blood are measured by biochemical tests from around ten weeks of pregnancy, which gives plenty of time to analyze cells by a definitive prenatal diagnostic test using amniocentesis or chorionic villus sampling. In the United Kingdom, AFP (alpha-fetoprotein) and HCG (a pregnancy hormone) are normally measured. Hospitals in the United States usually measure unconjugated estriol (another pregnancy hormone) in what is known as the Triple test; the more indicators used, the greater its reliability. In Leeds we measure all three plus another hormone (inhibin) to gain even more precision.

Scanning and blood tests can reassure worried parents, but they provide only statistical odds and no certainties. One in every 20 women who undergo the Triple test will be positive for an abnormality, but only one in 50 is eventually found to be carrying a Down's fetus. And a very small fraction of those who have a negative result and are given the all-clear are in fact carrying such a fetus (1 in 1,900). A positive result can cause plenty of needless anxiety, so the test is better reserved for pregnancies at high risk. All health screening is a gamble, whether for heritable disease, rubella, gynecological cancer, or pregnancy. How we weigh the risks and make our choices is subjective, as is any form of risk taking. Some people regard a 1 in 1,000 annual risk of dying from motor cycling as negligible; others forbid their teenage sons from riding. Likewise, some mothers regard a 1 in 600 risk for Down's syndrome as low; others are shocked and insist on further investigation. Each individual mother draws the line at a place that depends on her stake in the pregnancy and a host of other reasons that are difficult to discern, let alone to explain. Her final decision is something no one else should make for her.

Those opting for further investigation for Down's syndrome are in for a roller-coaster ride of emotions until the results of prenatal diagnosis are in. Other patients may be at lower risk for either Down's or neural tube defects, either because they are younger or have not had an affected fetus before, but are tested because they have a history of disease "in the blood." The risks of passing well-known diseases, such as cystic fibrosis, to their baby can be calculated by a genetic counselor in the light of a family history of the disease and from genetic tests of both partners. Prenatal diagnosis of a chromosome abnormality or a harmful mutation need not, of course, automatically imply abortion, but it can help to prepare the couple psychologically for a disabled child and alert the medical team to its arrival. What is more, it forces private attitudes and values to come to the surface.

In a recent survey by two "pro-choice" British charities, only two out of three people approved of abortion when the child was going to be born with either or both a physical and mental disability. One in five vigorously disapproved. We cannot agree what value to set on the life of a child, even a seriously defective one. Sometimes a single case becomes a cause célèbre, like the Baby Doe case in the United States. The baby was born with Down's syndrome and needed surgery to unblock a serious intestinal obstruction. When the doctors and parents decided not to operate, Doe died of starvation, triggering a national debate, though not the first of its kind.

Most of these tragedies are personal and private. One is movingly described in *Past Due* by the writer Ann Finger, who was herself disabled during childhood by polio and who once worked in an abortion clinic in California. She describes her own reactions when an obstetric disaster during a home delivery seriously damaged her child. No matter how carefully parents prepare to make everything safe and perfect for the child's arrival, lightning can still strike. The experience forced her to reassess where she would draw the line if decision time came around again. "It's true, I think I wouldn't abort if my child were going to have Down's syndrome or spina bifida. But if I were faced with the possibility of a child with Tay-Sachs (a progressive neurological disorder that causes death in early childhood) then I would have an abortion. I know, too, how easy it is to think 'I would do this, I wouldn't do that,' when you're not living the reality."

Possibly no country has a greater diversity of views on this subject than Israel. Abortion is anathema to the ultra-orthodox community, whereas many at the liberal end of society would opt for a termination even for minor defects, such as cleft lip and palate, which is reparable after birth. But the bias sometimes works in unexpected directions. It is now possible to screen for mutations that cause hardness of hearing or poor bone growth. Neither the deaf community nor people with a growth restriction regard themselves as inferior in any way, and some would prefer to have a child like themselves who would integrate more easily into their community. First reactions to the suggestion that a healthy child should be aborted in favor of a child with an impairment verge on outrage, but such a challenge to conventional values should at least make us question our own assumptions about disabled people as they struggle to overcome ancient prejudices.

Many people still do not want to play God and would rather leave any decision to a higher authority or to chance. Some feel that even the 1% to 2% risk of losing a healthy child after amniocentesis or chorionic villus sampling is too high. There are no easy answers, and these are circumstances when most people try to weigh the lesser of the evils. A decision whether to continue with the pregnancy of a defective child is perhaps the most difficult one that a couple ever has to face. In helping them make up their minds, the special expertise of genetic counselors is very important. Although they are trained to give advice "nondirectively," the final decision must be the couple's. There are no simple formulas that can help us answer what value we set on the life of another human being. And the dilemma will remain until the effectiveness of treatment matches our ability to diagnose problems that arise in the womb.

Social attitudes to abortion have a huge influence on the decision a couple takes, and one cannot help but comment on the striking polarity in attitudes. The official position of the Roman Catholic Church is that abortion is unconscionable except where the mother's life is endangered, a theological departure from St. Thomas Aquinas's doctrine of progressive development in the womb. Pope John Paul II upholds the hard line that "genetic screening is gravely opposed to the moral law when it is done with the thought of possibly inducing an abortion depending on the results. A diagnosis which shows the existence of a malformation or a hereditary illness must not be the equivalent of a death sentence." At the other extreme, American insurance companies are responsible for a countervailing coercive force. Some companies would like to wriggle out of any responsibility for medical coverage when they feel that not enough care was taken by parents to avoid the birth of a defective child. And in a bizarre turn in American courts, cases have even been brought against parents by their offspring for "wrongful birth." Surely some life is better than none, provided it is not blighted by pain.

An even darker scenario would be if governments used coercive methods to influence the judgments of individuals and their freedom to decide. In modern China, an official campaign is already underway to reduce the numbers of physically and mentally disabled people born. Like earlier campaigns, it is bound to fail, as already noted. Even the choice to opt for pregnancy screening services is best left to individual discretion, as such schemes carry with them tacit assumptions about our attitude to the unborn child. But it is important that individuals should be able to make informed choices, and it is in such an advisory capacity that the doctor plays a crucial part in guarding against abortion on trivial grounds. The Hippocratic oath may be an anachronism, but a code of medical ethics and peer pressure *should* provide much restraint, if not a completely uniform and watertight standard of advice and practice. This does not mean, however, that we should rely on the profession as a guardian of public morality any more than we rely on any other group in society.

The first diagnostic test to be developed to check for chromosome errors and mutations was amniocentesis. From about 14 weeks of pregnancy, a long needle can be passed into the watery chamber, or amniotic cavity, around the fetus to withdraw a few milliliters of fluid containing skin cells that have sloughed from its surfaces. An alternative method, called chorionic villus sampling or CVS, can be performed after 10 weeks. It involves taking a biopsy from the placenta for genetic analysis of the cells.

In the laboratory, the cells from these samples are cultured for two to three weeks until there are enough for counting the chromosomes accurately. Any departure from a normal set of 46 is serious, although much less so if there is just an extra sex chromosome than if any of the others are in excess. More subtle defects within a chromosome can be identified using special stains. But sometimes the slightest submicroscopic flaw in the enormously long DNA molecules can have the most devastating consequences. To identify such infinitesimally tiny irregularities, a method known as DNA amplification has been developed—"PCR" (polymerase chain reaction), in lab-speak. It involves making short molecules, or "primers," which are designed to attach to specific parts of the target gene. They are added to a tube containing DNA from the cell, together with a cocktail of molecular building blocks for making new DNA and enzymes to drive the reaction. Over a few hours and 20 or more cycles of heating and cooling to allow synthesis of new DNA strands and separation of the products, the single copy of DNA is multiplied at least a millionfold. Mutant and normal DNA sequences can then be distinguished by letting them separate in an electric field on a glass plate.

As the Human Genome Project advances, more and more genes come within the purview of screening methods based on DNA amplification. As soon as the molecular defects of genetic diseases are revealed, the possibility of testing prospective parents or fetuses or embryos is apparent. At first researchers focused attention on genes that have mutations with the most severe effects, like retinoblastoma and cystic fibrosis. They are called "penetrating" if they always produce an effect because they are not contingent on another factor. There is general recognition of the value of prenatal checking for these diseases, although if postnatal treatment becomes more effective, attitudes may change. Biologists are now finding increasing numbers of genes with mutations whose impact on health either does not take place until adulthood or depends on the effect of other genes or the environment and is therefore less than 100% penetrating. The question is now being asked whether conditions that may never develop in an individual's lifetime, or at least not until later, and that are sometimes medically treatable can be grounds for excluding a fetus and trying again for one that is risk free. One of the better known examples is Huntington's disease, which causes a fatal form of dementia and often at an age when people still have family responsibilities. There is also *APOE*, which is associated with a higher risk of Alzheimer's disease later in life. Perhaps best known is the tumor suppressor gene *BRCA1*, which carries an 80% lifetime risk of breast cancer and a risk of some other cancers in women and men. All these conditions have become relatively common in human populations since they do not strike until after the victims have transmitted them to their children and are not therefore screened out by natural selection.

Attitudes to "late" acting mutations are more mixed. For some parents, any risk is too high if they have a chance to "try again" for an unaffected baby. If they have witnessed the devastating effects of an inherited disease on another family member, it is understandable if they cannot bear to repeat the experience. Farther along the road of discovery, we may be able to screen for genes that have weak associations with conditions such as nearsightedness, eye and hair color and male-pattern balding. These are aesthetic preferences and no one would call them disabling, though we may prefer to avoid some of them for ourselves and our children. At one time myopia would have been a problem for hunters and a swarthy skin an advantage before people adopted clothing, but to choose them today would be only a trivial personal preference and there is no justification in using them to favor one or dismiss another fetus. If only "perfect" specimens were wanted, many of us, including myself, would have been counted out.

We should not simply assume that medical syndromes are always severe enough to warrant a death sentence in utero and must guard against simplistic genetic fatalism. Take Down's syndrome, which has become a byword for mental retardation and "uselessness." The great majority of people with the syndrome have an extra chromosome 21, but they are no more the same than all Englishmen are the same. Some never learn to speak, whereas others say their first word at 1 to 3 years old (compared to 1 to 2 years normally). A fair proportion go to school and eventually take a job. About 40% of children with Down's syndrome have heart defects, sometimes serious, some are hard of hearing or have thyroid problems, and most develop Alzheimer's disease in middle age. But a good number are healthy individuals who have rewarding family relationships. The Victorian doctor who originally described and dedicated his life to them, John Langdon Brown, would probably be horrified at our hasty dismissal of a Down's syndrome life. In 1866, he wrote: "They are cases which very much repay judicious treatment . . . they are humorous: they are usually able to speak . . . but may be improved very greatly by a well-directed scheme of tongue gymnastics." No wonder the Down's Syndrome Association does "not believe that having a baby with the condition is a reason to terminate a pregnancy." They do, however, accept that the decision is for individuals to make.

More research is needed to investigate whether it is possible to predict prenatally whether a fetus with three copies of chromosome 21 is more likely to have an IQ of 70 than of 30. Much progress is being made in new scanning techniques that provide an even more detailed view of the fetus. But our poor ability to predict a child's capacities at the time of birth should warn us not to place too much confidence simply in a better picture at an earlier stage of development. One day, we shall have technology that is more predictive than sheer chromosome number or a scan, so that the thumbs do not go down automatically when there is a Down's syndrome baby in prospect.

Many people would prefer to have a diagnosis much earlier in pregnancy to avoid the agony of having to decide about abortion when pregnancy is already well advanced. Doctors would prefer alternatives, too, especially if they avoided the risk posed by amniocentesis and CVS. Consider an imaginary scenario in a few years' time. A husband and wife are screened and each is found to be carrying a recessive mutation for Tay-Sachs disease. They want to know whether their fetus has inherited both copies of the defective gene and would become a child whose short life would be dogged by a major nervous disorder. The doctor takes a blood sample from the woman's arm at six weeks of pregnancy and delivers a diagnosis a couple of days later. There are grounds for hoping that such a blood test may arrive one day. Shortly after an embryo implants, a few blood cells start leaking across the placenta and into the maternal bloodstream. They are perhaps no more common than 1 in 10 million of the mother's blood cells, but that is enough for DNA amplification to pick up the gene or to diagnose whether the fetus is male or female. Each cell contains all the genetic information about the fetus we might ever want to know.

However, the method of fetal cell analysis is still beset by problems, and it will be a while before it becomes routine. Separating the cells from maternal blood is difficult, making it much more reliable to perform a test for a dominant mutation on the father's side than the mother's. Likewise, it is easier to identify a male fetus than a female, because a negative result for the male Y chromosome could mean either that a female was present or that the test had failed. Another problem is the persistence of fetal cells in the mother for months or even years after birth, which could make diagnosis for second and subsequent pregnancies impossible. Nevertheless, many of the technical problems are on the way to being solved. Using a cocktail of special hormones, a rare type of fetal cell, called a stem cell, might be encouraged to divide so rapidly that enough can be picked out from the mother's blood for study. And DNA fingerprinting can be used during the genetic analysis to confirm that the cell was from a new creature and not a carryover from a previous pregnancy or one of the mother's cells.

Whatever technology brings, we can be sure that couples will have more details on which to base their decisions in the future. But more information will not ease the anguish of having to decide whether or not to terminate a pregnancy. Consequently, the search has been on for ways of avoiding from the very beginning a pregnancy that might miscarry or be blighted by genetic disease. It has not been in vain.

Picking the Best

To increase the chances of successful reproduction by IVF, embryologists try to find the best eggs and sperm to make winning embryos. But choosing the best gametes is not easy, and so far it has proved more effective to make a selection at the embryo stage. Every day, embryologists in IVF clinics try to identify the best two or three embryos for transferring to the womb of the woman who supplied the eggs. This may sound like playing God, but why give the woman a poor or average embryo if she could do better? We always select the best apples from the produce bin, so why not choose the best embryos?

Embryologists are like a team of judges at a beauty contest, but when it comes to fine differences, no two have identical preferences. No one can finally be sure which embryo will have the best chance of succeeding in pregnancy. It makes sense to choose those with the plumpest and healthiest cells and to freeze the rest as backups. Those that have been fertilized with more than one sperm are discarded because they will have too many chromosomes, but that still leaves a lot of uncertainty. Some embryologists even try to improve the appearance of embryos by vacuuming out any dead cells with a fine pipette. It may or may not help the chances of pregnancy, but it is certainly the first cosmetic surgery any individual can have!

No matter how carefully the embryos are screened, every embryologist has stories about the ugly embryos that grew into the best-looking babies and the beautiful balls of cells that never made it. The judge that counts the most is the womb, and our poor ability to select the best has required the transfer of two or three embryos at a time (even more in some countries) to improve the chances of at least one of them being healthy enough to "take." Of course, if they are all good, it is likely that a multiple pregnancy will occur. The pregnancy may then be "selectively reduced," which is a euphemism for injecting a salt solution to kill the surplus embryos, leaving the one or two judged to be the "best." But apart from the occasional embryos that scanning reveals to be grossly abnormal or attached at the wrong place, such selection is often an arbitrary process. If embryologists were better at judging embryo quality, then only one embryo would need to be transferred and quandaries about multiple pregnancy—and reduction—would be avoided.

We might expect that the strongest embryos would have the biggest appetites. A nutritionist-turned-embryologist who works at York has managed to monitor nutrient intake by individual human embryos in vitro. Henry Leese's remarkable feat depends on ultrasensitive methods to measure nutrients in minute droplets of culture fluid. The weedy embryos did have smaller appetites, but genetic defects are not necessarily expressed at this stage and can easily go unnoticed. Perhaps being kept in vitro for longer, until such time that harmful genes might reveal themselves, would help to identify the ones best eliminated and improve the chances of pregnancy and a happy outcome. The results of research on this question are eagerly awaited.

It is much easier at present to detect genetic faults directly rather than waiting to see whether they affect the appearance or metabolism of embryos. The advent of DNA amplification has enabled embryologists to analyze a single cell removed as a biopsy from a human embryo. This method, which is called preimplantation genetic diagnosis, or PGD, was pioneered by Alan Handyside and Robert Winston in London to reduce the risk of implanting an embryo afflicted by a genetic disease. This is the most hi-tech way of choosing a baby. Though sometimes regarded as the back door to eugenics, it is really an extension of the subjective impressions that embryologists have been using for years to judge embryo quality. The difference is that PGD uses hard objective criteria based on abnormalities in single genes or chromosomes. At present, we can screen for only one or two defects at a time to avoid transferring affected embryos to the uterus, but the scope of PGD will be much greater as technology allows simultaneous analysis of many genes and chromosomes.

The first baby to be born after PGD, ten years ago, was in fact selected not after identifying a defective gene but because it possessed two X chromosomes and was therefore female. Detection of sex turned out to be an easier way of excluding the family disease, because it came only through the male line. PGD avoided the need to turn to gamete donors to avoid an affected son. In addition to diseases linked with sex—and there are about 400 of them, though most are very rare—tests are available for mutations on other chromosomes. These have to be detected by searching the DNA directly for the mutation. Research was focused at first on the most common one, cystic fibrosis, and tests for another dozen mutant genes have followed. There are 8,000 genetic diseases that have been recorded so far, and although all of them could eventually yield to analysis, many are extremely rare and occur sporadically rather than running through families.

Avoiding a pregnancy with embryos carrying a serious disease falls a long way short of the eugenic scenario that we have been warned about. A recent headline in *The Times* thundered "Designer Baby Enquiry Ordered," because the technique can be used to select embryos on the basis of a preferred sex, regardless of medical need. This choice is currently denied to patients under British law, though there is no evidence that any harm would be done if patients receiving PGD for, say cystic fibrosis, were to be given the choice of an unaffected boy or girl. The fear of choosing other personal preferences, such as a high IQ or a certain eye color, has been even more intense, though far less realistic. Unless a couple underwent IVF treatment repeatedly and built up a large stock of embryos for genetic screening, there would never be enough to find a specific characteristic unless it was common. Like a philatelist hoping for a valuable "mutant" stamp to appear in the daily mail, they would probably have to wait a long time.

The marvel of PGD is not that it offers unlimited choice but that the technology has been taken to the ultimate limit—the analysis of a single cell. Provided the cell is representative of the embryo and the biopsy does no harm, a complete inventory will in the future be drawn up just a few days after fertilization. Usually, one or two cells are removed with a fine pipette at about the eight-cell stage, like the embryonic equivalent of taking blood out of an arm. The scientist then makes the diagnosis in time to choose which embryos are safe to transfer to the womb the following day. From this stage onward, the course of events is similar to a normal pregnancy except that a CVS test is administered to confirm the accuracy of the original diagnosis.

The technology is getting faster and more accurate all the time. Chromosome abnormalities are being tested using a method called FISH, where F is for the fluorescent colored beacon which is used to label chromosomes for viewing under the microscope. For mutation analysis, the latest method in service is a fluorescent DNA amplification method using an automated gene scanner. It allows us to check more than one gene at a time and make a DNA fingerprint at the same time to verify that the cell was definitely from the embryo and not from a source of contamination. Already the list of diseases that can be screened in our laboratory has been extended to nine. The list of mutations that can be tested is lengthening rapidly, although it is progressively more difficult to perform tests simultaneously.

These methods that seem so sophisticated today will probably be regarded as primitive in a few years time. Mutation analysis is advancing rapidly to ever higher capacities and processing speeds—in much the same way as the computer chip. With DNA chip technology, arrays of DNA sequences representing hundreds or thousands of genes and mutations are laid out on a silicon wafer for screening.

When it comes over the horizon and into clinical practice, it will enable us to screen the rare mutations along with the common ones. Far from being just a gimmick, there is a pressing medical justification for the higher capacity. In the gene for cystic fibrosis, to name one example, there are at least 400 different mutations, and though in my area of northern England *Δ508* is the most common one, ignoring others is courting misdiagnosis.

I doubt that PGD will ever be used to screen every embryo created in IVF laboratories, although it will be more widely applied. So far, there have been only about 100 babies born worldwide after this testing has taken place, but eventually most couples who are already undergoing IVF treatment will want to weigh the costs and benefits of screening for common diseases, even if neither partner has a family at special risk. In northern Europe, 1 in 20 people carries a mutation for cystic fibrosis, and in southern Europe, beta-thalassemia is even more common. In Ashkenazi Jews the major concern is Tay-Sachs disease, and in African Americans it is sickle cell disease.

The impact of PGD on the frequency of recessive mutations occurring in the population is negligible, but the benefit to a family of avoiding a member being affected is incalculable. With dominant mutations, however, PGD and prenatal diagnosis would reduce the frequency of disease appreciably but for the fear of testing. Huntington's disease is a good example since it is due to a highly penetrating mutation and probably arose in one or perhaps just a few individuals long ago and was passed down through their families. But the dread of the disease is so great that many prospective parents do not want their embryos or fetus to be screened, because if the test does prove positive, the knowledge that one of them is definitely carrying a mutant gene for a condition that progresses inexorably is almost too much to bear.

Among the other people who avail themselves of the opportunity to screen their embryos will be those whose eggs are suspected to be of poor quality and who have had repeated miscarriages. Excluding embryos with abnormal chromosome sets could improve pregnancy rates and diminish the need for egg donation. Similarly, many members of the baby boom generation who have postponed their families until they are over 35 are now turning to ART for help. Screening embryos for chromosome 21 could give them greater peace of mind from the very start. Younger patients would benefit less. Nevertheless, since chromosome abnormalities account for many of the miscarriages that occur after IVF, they might find that the financial advantages of using PGD outweigh the costs of repeating a failed IVF cycle.

Screening the egg for Down's syndrome is emerging as an interesting alternative to screening embryos, especially since problems are less often due to the sperm. Eggs can be tested by skillfully removing their "polar bodies" as soon as they have been collected from the ovaries and before they are placed with the sperm. The polar body is a tiny bleb that is thrown out by the relatively huge egg, rather like the moon's inception when it was thrown out from the earth. The polar body is doomed to degenerate, but it contains a surplus set of chromosomes that are available for screening. Most chromosome errors in older eggs occur shortly before ovulation, and the chromosome tally in the polar body gives a strong indication of the numbers in the egg. If the polar body is missing a chromosome 21, then the egg probably had more chromosomes than it should have and is best not used. The polar body could even be used to screen for single gene defects on the mother's side, although the vagaries of biology make the results less certain.

Fresh mutations crop up all the time and are more common in men than women—especially men over the age of 50. A mutation cropped up in the testes of the 51-year-old Edward, Duke of Kent, and male descendants inheriting it through his daughter, Queen Victoria, were affected by hemophilia. We might expect that the sperm that gets to the egg first—out of 200 million ejaculated at one time—would be healthiest. But many a genetic defect has been carried in a champion sperm, and on the other hand, the ICSI technique has produced many normal babies from crippled sperm. Sperm—for which there is no equivalent of the polar body analysis—rarely show their true colors until their genes are expressed some time after the embryo has formed. Apart from performing tests on the man himself, which even then cannot reveal a random fault in a sperm cell, eggs will continue to be at risk from dangerous liaisons.

A man carrying a harmful dominant mutation should consider other options for having a child, such as using a healthy sperm donor or adoption. There are special cases, however, where firing his own sperm is a game which, like Russian roulette, is inadvisable but can sometimes be won. A man who develops Huntington's disease at, say, age 50 is likely to have a son affected at 30 and a grandchild who dies of the disease in teenage years. This deadly progression is due to the copying of a few DNA sequences with the code CAG when the sperm are forming. This stuttering mutation produces no neurological symptoms in the man unless it has been repeated at least 36 times. But a man at this threshold will perhaps have only one sperm with that number of repeats out of every hundred with more, which accounts for more pronounced effects in the next generation. There is a chance of avoiding transmitting the disease with his own sperm, but it is a slim one, and science is still helpless to identify a safe cell for him to use.

PGD might appeal to more of the fertile population at large if embryos could be tested without the trouble and cost of IVF. If embryos could be flushed from the womb at the free-floating stage a few days after fertilization, they could be screened so that the couple could decide whether to proceed with the pregnancy or not. PGD would then be reduced to an "office procedure" without the need for a general anesthetic or invasive surgery. Embryos would wait in culture fluids overnight until the genetic analysis was complete and a decision about which one to implant could be made. This is not such a fictitious scenario as it might seem: it is now routine in farming to recover embryos from cattle of high genetic worth for transfer to surrogates of lower commercial value. The numbers of embryos are increased artificially by injecting fertility hormones to superovulate the cows, and the procedure can be repeated at regular intervals. Is this the future for humans too?

In California 15 years ago some doctors used a fine tube passed through the cervix to flush out human embryos from the wombs of consenting women. The aim was to obtain embryos for donation to other patients and, although this technique was quickly superseded by egg donation, it might one day be perfected for routine PGD. But progress has been slow because there is no foolproof method of collecting all the embryos and avoiding accidentally forcing them back up the Fallopian tubes where they could produce an ectopic pregnancy. In effect, hi-tech laboratory science has forged far ahead of our doctors' ability to put such techniques into clinical practice. That is a pity for people who wait to undergo PGD or cannot afford it as it presently stands. Those who fear that selection of embryos will take us further down a slippery slope should remember that there is no equality of access to fertility services in most countries and that PGD was developed to avoid the angst of prenatal diagnosis and late abortion. Besides, it is far less breathtaking than the next step forward.

Hitting the Bull's-Eye

I regard it as self-evident that it is better to heal than to kill. It is a natural corollary of the physician's code, *primum non nocere*: do no harm. Yet attempts to modify the genome in gametes and embryos to avoid disease are widely regarded as abhorrent, whereas abortion is tolerated. This so-called germ line therapy has already been prohibited in many countries before there is any chance of putting it into practice. There is perhaps no technology—real or imaginary—

apart from cloning, that has come in for more criticism than influencing a baby's genetic inheritance. Yet that is precisely what we do in selecting a mate and his or her genes—and that has always been regarded as natural.

When a safe technology arrives it will be vexing if we are prohibited from trying to heal defective embryos and fetuses and told simply to dispose of them. Imagine the reaction there would be if organ transplantation were prohibited because it is "unnatural"—though that is what some people called for when transplantation was a medical novelty. It is hard to see how the replacement of a defective gene is any less "natural" than the replacement of a defective organ. Indeed, the major difference is the entirely beneficial one that medical intervention need occur only once around the time of conception, and the benefits would be inherited by the child and its descendants.

It goes without saying that germ line therapy should not be tested in humans until we are sure it will do more good than harm, and it may be a decade or so before we can be satisfied on that score. In any case, it is necessary to prove that treatment of diseases in existing patients—both children and adults—is effective before any serious efforts can be made to intervene at such an early stage in the reproductive process. Treatment after birth is called somatic cell therapy because the specific cells affected by the disease—such as blood, muscle, and brain—are targeted. Such treatment is not intended to affect the germ cells or to pass on the supposed benefit to a patient's offspring, though this might happen inadvertently. Gene therapy trials have so far produced generally disappointing results, and it has proved more difficult to get DNA stably integrated into human cells than originally anticipated. But eventually the research endeavor will pay off and gene therapy will be available. This worries some people, although there is nothing in the genome that is pure and perfect nor such a thing as a standard genome. There are as many genomes as there are unique human beings, although all of our genes have been written in the same four-letter code like an

alphabet. Unlike the Bible, which has been carefully transcribed by generations of scribes and theologians to avoid
corrupting a holy text, the human genome is ever in flux.
The numbers and types of genes have changed over the
long course of evolutionary history, although the DNA foot-
prints of our ancient ancestors still leave a prominent
mark. Many of our genes started out in cells like yeast and
passed through trilobites and other long-extinct animals.
Only the genetic novelties that helped to make thriving offspring have been preserved and last to this day.

The aim of germ line therapy is to preserve the genetic
legacy that has served us so well. If that implies replacing
the defective part of a gene with the correct human DNA
sequence, so be it. If a gene from a fish or a fly is helpful
in combating an illness or saving a life, why not use it?
Most of our human genes have been derived from the same
ancestral stock, and there are plenty of precedents for
using animal products in human medicine: some domestic
animals, for example, provide us with serum to vaccinate
against parasites, others their pancreas to prepare insulin
for diabetics.

It should also be noted that "genome intrusion" occurs
naturally from time to time. A very long time ago, a free-liv-
ing bacterium-like organism entered a cell and changed the
course of biological history. Both the host and the tenant
liked the arrangement so much that they made it permanent and set the pattern for the evolution of more complex
cells. The bacteria became mitochondria, which are little
organs supplying energy for the whole cell. Mitochondria
still have their own strands of DNA but now need genes in
the "host" cell nucleus and can no longer live apart. Most
plants too have benefited from mitochondria, but they also
acquired another organelle from ancestral blue-green
algae. This became a chloroplast, which made sugars and
gave plants their color. The advantages of importing—or
transfecting—foreign DNA are now all around and within us.

DNA can also pass between individuals of the same
species, and not only during intercourse between opposite

sexes. Bacteria acquiring beneficial mutations that give them resistance to antibiotic attack can pass the gene from one to another in the form of little rings of DNA called plasmids. The recipients pass them on to their progeny, which flourish all the more. The advantage of gene transfer was discovered by bacteria long before any scientist dreamed of it.

We are less sure how open the human genome is to foreign DNA, though there are indications that importation has happened in the past, even in addition to the handful of mitochondrial genes in our cells. There is a family of genes called tumor suppressors which, when mutated, are no longer able to guard the integrity of the cell and which may become cancerous. One of these suppressors is the previously mentioned *BRCA1*, and if it becomes defective it can encourage breast and other cancer to develop at a young age. Reading the sequences of the unmutated genes, biologists found evidence of molecular signatures that suggests that they may have been carried over from a bacterial or yeast-like infection eons ago. The modified ancestral cell has passed the survival advantage down to the present day. We had assumed that genes are transmitted only vertically from father and mother to children, but tumor suppressors seem to be examples of horizontal DNA transfer between cells, which is the same strategy as germ line therapy.

Transfection and integration of foreign DNA into cells may still be occurring, if only rarely and at considerable risk. A balance has been struck between the benefits of a genome stable enough to resist harmful changes and one flexible enough to respond to advantageous ones. The discovery of "jumping genes" in animals transformed scientists' view of the genome, which now seems more plastic and amenable to experimentation than we had imagined. About a third of the human genome consists of "mobile elements" in the DNA; they are kept under control by chemical methylation to avoid chunks of DNA randomly transposing themselves, and sometimes by chance getting into vital genes. Just occasionally a transposition occurs nonetheless, and the effect is then like taking a chapter from the Old Testament book of Daniel and slotting it into Revelation in the New Testament. Some of the big jumps in evolution may have occurred in this way, explaining the discontinuous fossil record, which Steve Gould has called "punctuated."

As far as germ line therapy for curing disease is concerned, we have to grapple with the practical question of how to introduce new DNA into the cell nucleus. Retroviruses have no trouble doing so because they are infectious agents like HIV; they enter cells readily and hijack their machinery to make more copies of themselves. Their genes are made of RNA instead of DNA, but it is converted by a special enzyme into DNA, which leaves molecular fingerprints behind in the nucleus. The change in the host's DNA can trigger malignant disease, especially if a tumor suppressor gene has been affected. But understanding how retroviruses manage to trick the cell may not only help to prevent or cure AIDS. It may help us to learn how nonpathogenic viruses might be used as Trojan horses to carry human DNA into cells to replace defective genes, such as those that cause cystic fibrosis.

One of the causes of optimism is that there are many potential options for correcting or replacing a faulty gene, and molecular biologists have been very inventive. Instead of replacing an existing gene, it is possible to make an antigene, which is a DNA sequence that can make RNA molecules for knocking out the products of a mutant gene or damping excessive gene action when an extra chromosome is present. The agents for these actions are pseudogenes, which are found naturally in the genome, and artificial human chromosomes, which are currently under development. Introduction into the large egg cell would be analogous to what happens naturally during the transfer between bacteria of plasmid DNA. It is perhaps too wild to imagine that abnormal gene activity in Down's syndrome could be damped down so easily before the fetus was harmed, but dreams sometimes come true in science. We can expect to see steady progress as the dawn of a new technology breaks, enabling us to correct faulty genes and rescue precious embryos from genetic disaster.

Perhaps it is better to avoid such a delicate molecular balancing act and try a different tactic. Plucking the harmful extra chromosome 21 from a fertilized egg currently seems an impossible feat, but with the advance of microscopic methods it may come within our reach. Since this act would cure the embryo by removing the offending chromosome, this form of germ line surgery must surely be acceptable. Overcoming the effects of mutations in the DNA molecule requires chemical therapy, but instead of tampering with the gene itself, it may be safer and more efficient to use a molecular toolkit called a ribozyme. This is a small molecule with natural enzyme activity that can carry out repair work on the RNA product of the gene without altering the original flaw in the DNA. Only the sweat of laboratory research can reveal which of these techniques will be medically effective.

Meanwhile, scientists are still looking for a "magic bullet" that can fire the DNA into its target without also damaging other parts of the genome. Currently, we are more like a blind man shooting a blunderbuss—more likely to fill the bird with shot than make a "clean" killing with a single pellet. There are many dangers in trying to repair a gene or introduce extra genes, and if a mistake is made in a germ cell it could be inherited and blight the lives of future generations in the family, which is why germ line therapy must be considered even more carefully than somatic cell therapy. Although so far it has proved difficult to deliver DNA accurately into human cells, the "bullet" approach has had some remarkable successes in animals. Only 15 or so years ago the genetic modification of animals by injecting DNA into a fertilized egg was considered sensational. There were newspaper headlines about giant mice with extra growth hormone and a sheep that produced a human clotting factor in its milk after injection with fertilized eggs with DNA encoding the respective genes. But it was a chancy method—sometimes there were many DNA copies, but 98 times out of every 100 there were none. The high frequency of fetal deaths and congenital abnormalities caused by the experiments was also worrying. Other methods were tested, including hitching foreign DNA to a fertilizing sperm, but the results were unpredictable. This disappointment is not so surprising in hindsight, for if sperm were able to act as easily as carriers, any fragments of DNA floating about in a woman's fallopian tube might be drawn into the egg during the act of fertilization.

Nowadays, extremely precise genetic changes can be made in the laboratory. Every week we hear about mice that have been given a gene before birth to replace a defective one or have had another one knocked out by a mutation to see what will happen. Many people have heard about the famous "oncomouse" that develops cancer after an induced genetic change, though most heritable changes are not so adverse. Immense benefits are flowing from this technology as it provides a better understanding of how diseases develop and can best be treated. When the muscular dystrophy gene was knocked out in a mouse, the cells soon switched on a related gene called utrophin. If this backup gene could be turned on in human patients with the disease, much anguish after prenatal diagnosis could be avoided. Other experiments with mice have shown that it is possible to reverse a natural mutation by introducing a correct copy of the genetic flaw to make the mouse completely well. There are many such revealing examples of genetic engineering in animals.

If these changes can be made so accurately in a mouse, why not in a human too? Hundreds of strains of genetically modified mice have been produced, some of which have exactly the kind of prescribed changes that could help humans overcome a disease. The process of targeting a short segment of DNA to exactly the right place in the genome so that the corresponding part can be exchanged is known by the ponderous term "homologous recombination." It is a rare event. If we transfected 10,000 cells with DNA we could expect only one of them to fulfill this stringent requirement. Not only is the method too wasteful to be used with human cells, it is also completely impractical. The methods used for making these mice are indirect and require passaging the cells through host embryos and selective breeding. Quite apart from the ethical minefields we would have to tiptoe through, the desired genetic change in a human's gametes would not appear in full form until his or her great-grandchildren were born.

These methods work well in mice but not in other species, including farm animals. That is why the new cloning methods are so important for making genetically engineered sheep and cattle. The breakthrough came in Scotland, most memorably with the birth of Dolly, the cloned sheep, which is described in Chapter 5 along with the implications for making human babies. Dolly was made without any genetic modification, although another animal in the same flock, Polly, had DNA for human clotting factor IX introduced when she was cloned. It is hoped that she will one day produce enough of the factor in her milk that it can be extracted for patients with hemophilia type B. The advantage of cloning is that somatic cells from an embryo or a fetus or even an adult can be grown in large numbers and the few that are successfully transfected with DNA can then be fused with eggs. Those that make it to the embryo stage are transferred to surrogate mothers to develop into a lamb like Polly. Once she delivers her first lamb, the company that sponsored the research hopes to be milking the profits.

Cloning provides a way in which inherited diseases can theoretically be corrected in humans. But need treatment be this complex? There is always the danger that if we overcook we spoil the dish, and this is presumably as true in human reproduction as anything else. Besides, we shall meet later some serious practical obstacles to human cloning. But before moving on to them, it is worth noting that cloning hammers one more nail into the coffin of the notion that germ cells should remain genetically inviolate. This idea harks back to the end of the nineteenth century when the separate development and lineages of germ cells and somatic cells were stressed. Like the old monarchies of Europe, where fraternizing with the proletariat was discouraged and royal blood was kept pure and aloof (or as much so as human nature could restrain itself), these two cell types were considered to be separate entities. But Dolly and Polly were made from an egg in which the sperm was replaced by a somatic cell. In other words, a somatic cell can serve as a germ cell, and there is no logical basis for drawing a sharp line between therapies using somatic or germ cells.

In the course of evolution a lot of genetic novelty is thought to have come through mutations introduced by the sperm. We should consider therefore whether a genetic disorder could be treated in the testes. The male organs have the advantage over the ovaries in continually producing new gametes from stem cells. If the stem cells could be grown in vitro, there is a chance that some of them might integrate desirable genetic changes after transfection with DNA. While we await news of this experiment, scientists at the University of Pennsylvania have shown that stem cells recovered from mice can be transplanted back into the testes to make sperm again. The project for making a correction in the male germ line is therefore at the halfway stage. My colleagues recently embarked on stem cell recovery from the testes of young men who are undergoing chemotherapy for Hodgkin's disease, which may sterilize them. Normally, these patients bank frozen semen specimens, but this safeguard is not always reliable, and the possibility of restoring natural fertility by stem cell transfer is appealing. Besides, stem cell storage is the only way we might be able to preserve fertility in child patients in the future. It is now routine to bank bone marrow cells for transplantation after chemotherapy to save a life, and it is natural to want to save fertility by storing cells from the testes. We never considered genetically tampering with the cells of our patients, but gene therapy might be an option one day for men with inherited diseases if we can succeed with transplantation and restore fertility.

Stem cell therapy addresses problems in only one sex, of course, and matters in genetic health and disease are seldom equal. We used to think that it was immaterial whether a mutant gene was inherited from the father or the mother, but now we know that a few genes "remember" where they came from, and this makes a difference to the way they act in the embryo and later in life. Taking the example of the Prader-Willi syndrome, a condition that involves mild mental retardation, food obsession, and obesity, a gene defect in chromosome 15 is responsible, but the disease becomes manifest only if it is inherited from the father's side. The gene on the other chromosome 15, inherited from the mother, is said to be "imprinted" because it is inactive as a result of methylation.

Now consider Angelman syndrome, a condition that causes a greater degree of mental retardation, speech difficulties, and epilepsy. This too is due to a mutation in a gene on chromosome 15, but in this case the mother is the origin of the defect, and the father's gene is imprinted. Since the opposite gene in either syndrome may be normal, it is theoretically possible to call it into action by demethylation to avoid the disease altogether. We know of only a few imprinted genes, but there are quite possibly many other mental diseases that share this characteristic.

Angelman syndrome has been linked with autism. Occasionally an autistic patient is discovered who, although perhaps having a low IQ and typical autistic features, has such an amazing talent that we are stupefied. So-called idiots savants have extraordinary capacities for memorizing calendars and musical scores and creating wonderful works of art and music. No one knows much at all about the genetic basis of autism. But if the talents of such people flow from normally untapped potential within the human brain, we may speculate whether this is a result of different patterns of expressed genes. Could some of these genes be imprinted and lying silent in normal genomes? This might explain why geniuses usually turn up completely out of the blue. If so, perhaps a future generation of researchers could discover ways of releasing the intellectual and artistic talents of people by influencing the expression of the genes they were born with. At present such a thought is little more than a fantasy, but it raises the question of the extent to which we might be able to *enhance* human beings by genetic engineering.

Gilding the Lily

Couples planning to conceive take more care now than ever before and are offered plenty of advice about prudent diets, adequate exercise, and the dangers of drugs, smoking, and alcohol to an unborn child. Even some men are taking notice at last, modifying their recreational habits and loosening their jeans to give their more modest biological endowments—sperm—the best chance. Only by bringing together the finest ingredients can you make a classic dish.

The Ancient Greeks advised pregnant women to gaze at statues of Castor and Pollux and other beautiful figures to make their children fair and graceful. Two thousand years later, French-women visited the Louvre for the same reasons and with the same hopes. It is said that Anita Roddick, the famous entrepreneur and business guru, played Beethoven through a microphone on her body to her unborn daughter. There is no evidence that aesthetic experiences during maternity can help the fetus, but neither is there any to the contrary, and they cannot do any harm.

A biblical precedent for transferring desirable impressions from mother to offspring was set by Jacob. His wily Uncle Laban tricked him into serving as a herdsman for 14 years to earn the hands of his two daughters. Later he broke the promise he had made with Jacob, who was due to receive all the lambs of impure color bred from the flocks so that he could build up capital for his own family. But Jacob was more cunning. He fashioned wooden rods with stripes of peeled bark and set them at a watering hole. When the stronger ewes and nanny-goats gathered to mate there, they came under the influence of the patterns they saw and "brought forth striped, speckled, and spotted (lambs)."

Jacob's experiment was a sort of scriptural precedent for Jean-Baptiste Lamarck's optimistic theory of evolution. The great French biologist of the early nineteenth century thought that children inherited the characteristics that, for good or ill, their parents had acquired during life. He would have said that giraffes had developed long necks because their parents had to stretch to graze high in the trees. Likewise, the blacksmith's son inherits his father's strong arms and a daughter acquires her mother's skills at the piano keyboard. Lamarckism has a seductive appeal, as it seems to assure us of reward for diligence. It was widely accepted before Darwin came along, and even long after it had been scientifically discredited, Trofim Lysenko was still applying it to Soviet agriculture—with disastrous consequences.

The application of Lamarckian theory to childbearing has been equally negative. Down the ages, women have been blamed and punished for giving birth to deformed babies, their behavior during pregnancy deemed responsible when no other explanation was apparent. Bad temper or melancholy were said to risk the health of the unborn child. Even the sight of a hare was thought to produce a child with a harelip. Pregnant women were warned never to look at a deformed person. The Siamese twins Chang and Eng were forbidden to perform their acrobatic acts in Paris for fear of the danger that the sight of them would pose to pregnant women in the audience. The poor Elephant Man, Joseph Merrick, held such maternal impressions to be responsible for his own pitiful condition: "My mother was going along the street when a procession of animals was going by. There was a terrible crush of people to see them, and unfortunately she was pushed under the elephant's feet, which frightened her very much; this occurring during a time of pregnancy was the cause of my deformity."

Were any of these stories true, there would surely also be some signs that children inherited resistance to epidemic diseases from their mothers, because the survival advantage would be expected to spread quickly through the population. But convincing evidence of inherited immunity after parental immunization or exposure to smallpox, diphtheria, or any other infectious disease has never been produced. There is an infusion of maternal antibodies across the placenta and in the colostrum of the milk that affords some temporary protection, but this is nonheritable. If the biological bonus of acquired immunity is foregone, it is even less likely that parents would biologically transmit physical strength or technical or mental skills to their children. Lamarckism is dead, although the cultural inheritance of know-how thrives, of course.

The modern dream of improving nature is now based solidly on molecular genetics and reproductive biology. In contradiction to the generally pessimistic tone of most newspaper headlines announcing breakthroughs in these fields, many members of the public are surprisingly accepting of new technological ideas. In a recent Gallup survey of attitudes about the use of genetic engineering for a British newspaper, 80% of people approved of using genetic engineering to reduce cancer risk. A still sizeable percentage also agreed with reducing human aggression and improving eyesight, even though these can be treated in other ways. Men were generally more enthusiastic about vain exploits than were women; half of them were in favor of improving intelligence and a quarter thought that enhancing female attractiveness was a good thing, though the survey report did not relate how.

The declarations people make when stopped on the street must be taken with a pinch of salt, but there is no doubt that a sense of genetic determinism is pervasive in modern society. Almost everything is assumed to have a genetic cause and hence a genetic remedy. Whenever an interesting new gene is discovered, pundits on television and newspaper columnists speculate wildly about how the knowledge might be applied, sometimes revealing a degree of faith in our genetic destiny that science does not warrant.

There has been plenty of heavyweight criticism to counter some of these assumptions, but many people can still be easily persuaded that the gene for aggression is on the Y chromosome and the gene for shopping is on the X (women have two copies)! Underlying much of this thinking is a sort of biological Calvinism that reached its nadir in the past when it attributed mental caliber to particular classes and races. The modern oversimplification about the inheritance of IQ goes something like this:

- There is evidence that IQ is heritable.
- Heredity is based on genes.
- There must be a gene for intelligence. QED.

Geneticists have been at pains to deride old attitudes about nature versus nurture, regarding the two as intimately connected. Genes never act on their own. Although knocking out a specific gene in a mouse has sometimes produced a very precise and predictable effect, more often there have been multiple effects, and occasionally there have been none at all. There is rarely a simple relationship between genes and the phenotype. Genes are never *for* something, though they may be associated *with* something. They act on a specific part of the body structure or metabolism or are involved in chemical signals, all of which interact with marvelous complexity in the cell. The role of genes in the physical and mental development of living creatures, as well as in many diseases affecting them, is far from simple. The primary genetic lesion may be clear-cut, but the effects downstream in the cell and on the physiology of the whole body are often manifold. Every time you read the word "syndrome" for a medical condition, you can be sure that the disease has many manifestations, which geneticists call "pleiotropy." These manifestations reflect the diverse effects of the gene in the cell where it is expressed, probably with effects echoing around the body.

We must bear all this in mind in considering the prospects for genetically enhancing the human body. What constitutes enhancement is subjective, but perhaps people most readily think of a greater endowment of physical beauty, athleticism, and, above all, intelligence. It is disappointing that we should seek the characteristics that often give power and influence over others rather than those that make for greater cooperation and bonding between people. But all of them are complex and difficult to define and so probably the least amenable to enhancement.

I have chosen beauty as the first of my "three graces," because a pleasing appearance matters in society, perhaps more now than ever before. A beautiful baby is thought to be a fine reflection of its parents' qualities and has a better head start in life, though it may not be any happier. Ideas of beauty vary from culture to culture, and any attempt to establish a common standard is doomed to failure. As the Scottish philosopher David Hume observed, "Beauty in things exists in the mind which contemplates them."

So there is little prospect that genetics could be used cosmetically to improve human features because we just wouldn't know where to begin. The face is shaped in the fetus by the movement of sheets of cells. The signals that cause this movement come not only from genes in the sheets of cells themselves but also from neighboring ones and others further afield. What seems like a simple job of molding a lump of clay is in fact very complex, though a few of the key genes, such as the one called *sonic hedgehog*, have been identified. No Dorian Gray gene for handsomeness, this gene is also involved in the formation of parts of the brain; any genetic tampering, except to undo a harmful mutation, is therefore risky. Even features as apparently "simple" as hair color have turned out to be complicated. For example, studies of mice suggest that there are as many as 100 genes involved in the pigmentation of their coats, and there could be as many—or more—in humans.

My second "grace" is athleticism, because this is something to which in today's sports-obsessed society so many people seem to aspire. The lineup in an Olympic sprinting event suggests that some races are better equipped biologically for speed than others, although we are still a long way from finding any genetic explanations for differences, nor do we know on what genetic criteria to select the best runners or how to make them go even faster. Clearly, there must be many genes involved in shaping the physique and enhancing physical stamina. Certain variants, or polymorphisms, of a gene called angiotensin-converting enzyme, or ACE, have been found to be associated with elite performance among high-altitude mountaineers. They seem to make the muscles more tolerant to prolonged activity, perhaps by improving the circulation, but whether they are as good for a healthy heart in middle age we have yet to discover.

The final "grace" is intelligence, because this is the distinguishing and perhaps most desired feature of our species. When young mothers were surveyed they put "good brains" at the top of the list of wants for their baby, and fathers are equally keen for their offspring to be able to successfully jump the academic hurdles en route to a prestigious or well-paid career.

The human brain is by far the most complex organ in our body, and it expresses more genes than any other. It would be surprising if inheritance played no role in determining its higher cognitive functions, and the search is on for genes associated with superintelligence in young people whose IQ exceeds 170. One of the principal behavioral geneticists in the field, Robert Plomin, an American based in London, is convinced that IQ—which is the token we take for intelligence—is one of the most heritable aspects of behavior. There is a great deal of controversy about the precise degree to which IQ is inherited. Comparisons have often been made of identical twins growing up together or apart. The famous British studies of twins done many years ago by Cyril Burt, which purported to prove high heritability, have now been discredited as biased and based on falsified data.

Heritability is usually ranked on a scale from 0 (all environmental and no genetic effects) to 1 (all genetic and no environmental effects), and estimates for IQ have varied from 0.3 to 0.6 or even wider. Taking these figures as examples and making a few simple calculations, we can quickly reveal the futility of any speculative attempts to screen embryos or transfect them with DNA from genes associated with bright children. A heritability of 0.3 does not necessarily mean that a full 30% of IQ is due purely to genes, for much could still largely depend on interactions with the environment. This is clear from the greater variability of IQ scores in twins raised apart than in twins who were always together. If we take for granted the fact that the high heritability of adult height is profoundly influenced by the environment (diet), we should willingly accept the same conclusion for intelligence. But assume, for the sake of argument, a figure of 0.3 and that a gene is found that can account for 5%— which would indeed be a major gene. What is the net benefit to the individual? Its presence would then contribute just 1.5% to the total scale, and it would boost the IQ of an average person from 100 to only 101.5. Nor would a heritability of 0.6 make a lot of difference, for although the value is doubled, the IQ score would still rise by no more than 3%. These are trivial advantages, and it would be better to invest the money in education.

Every caring parent wants to be proud of sons and daughters, to see them enjoy full health and succeed in whatever walk of life they choose. Science has become very effective at screening for the genetic seeds of disease at the earliest possible stages, and the scope is ever broadening. Although it may be possible one day to correct certain errors in a precious embryo, improving on a healthy blueprint for physical and mental characteristics still looks like a futile quest as far ahead as we can foresee. In addition to the biological arguments, there are social ones. We can have no confidence, as yet, that selecting genes that are more frequent in superintelligent people will help anyone to become a better banker, teacher, or football player. Society rewards success, confers privilege, and enables people to rise to the top in ways that owe very little to genetic endowment. In England, all the present High Court judges were educated at either Oxford or Cambridge, and most of them went to famous schools, like Eton and Harrow. The social factors in their success are obviously preeminent.

These conclusions should be reassuring. It would be hard if the whole of our lives really were mapped out in our genomes. Science give us some hope—but no more than that—of triumphing over the biological flaws at the beginning of life. There will still be some surprises on the day the baby arrives for a long time to come.

5

Keep Out the Clones

Hello Dolly

The road that led to Dolly, the cloned sheep, began in a pub. A chance conversation with another scientist convinced Ian Wilmut that farm animals could be cloned by fusing cells from embryos and fetuses. Not many scientists at the time were brazen enough to try. Fewer still dared to ask whether adult cells might be cloned. And so, with few thoughts about the controversy he would cause, he quietly laid plans for some of the most sensational experiments of the 1990s.

Cloning was not developed to produce identical flocks or with a long-range aim of making copies of human beings because the conventional way of making babies is still the best. Ian hoped to use the technique to genetically engineer a few sheep so that they could make some medically important products in their milk; he planned to then use natural breeding to increase their numbers. A few years earlier at the same research station near Edinburgh, his colleagues had succeeded in making a sheep with human DNA for a blood-clotting factor. When the factor was produced in the udder and extracted from the milk it was useful for treating patients with hemophilia.

The transgenic animal was made by injecting DNA into a fertilized egg, but it was a haphazard method that was hard to repeat. Ian realized that it was better to target a nonreproductive cell from an embryo or a fetus or even from an adult udder. Methods were already available for introducing stable and heritable changes into the DNA of somatic cells, and, because they can divide to make millions in vitro, it is possible to select a cell with the perfect modification. After fusing the cell to an egg whose own genetic material had been removed, the animal would produce a valuable product and pass this virtue on to its lambs. Half the world now knows that he was successful.

The team was anxious to avoid criticism about the provenance of the lambs—skeptics would be bound to suggest that they had been sired after a sneaky natural mating with a ram and not cloned after all. They therefore arranged to use different breeds of sheep for donors and recipients. Welsh mountain sheep provided the donor nuclei, and Scottish Blackface sheep were used as egg donors and surrogates. If cloning were successful, the lambs would have a white face. They did, and DNA fingerprinting provided the final confirmation.

The experiments were done with a modest outlay and skills that any large IVF clinic can command, which makes cloning humans seem both more likely and more scary. Once the eggs had been collected from sheep's ovaries and placed in a petri dish, their chromosomes were removed with a fine pipette. The donor cells were prepared in a special medium and injected under the membrane of the egg using the deft movements of a micromanipulator instrument. Then a brief pulse of electricity was passed to fuse the mass into one and restore the normal number of chromosomes without the need for a sperm. At the same time it activated the egg, which started dividing to form an embryo. These "sparks" used to start life harked back to the manner in which Frankenstein had animated his monster and were greeted with glee by some journalists. But Ian Wilmut's gentle manner and sage appearance spared him from being cast as the fiendish scientist.

Before long, flocks derived from cloned parents may well be commonplace, and the sheepdogs guarding them could be clones too. The commercial incentive to clone domestic animals, from racehorses to family pets, will drive the technology faster forward than will any medical imperative. Imagine a vet some time in the future taking a tissue biopsy or a blood sample from a dog or cat. He would make a genetic likeness by fusing the cells with eggs extracted from the ovaries of young bitches or kittens undergoing sterilization. The cloned embryos could be transferred to the wombs of the same pet, or more likely to a surrogate chosen at the pound.

When the first cloned sheep were announced by the Scottish scientists in 1996, there was merely a murmur of publicity, and the arrival of the cloned calves, Marguerite and Mr. Jefferson in France and America, drew only moderate interest. Yet they signaled a revolutionary breakthrough and the promise of pharmaceutical riches in milk when the next crop of clones was designed with DNA for precious medical products. These animals were cloned from embryonic or fetal cells. It took the cloning of an udder cell from a 6-year-old ewe and the birth of Dolly to start the uproar. The potential medical benefits of cloning animals were set aside and attention was firmly fixed on the production of identical copies of *people*. There was also wild speculation about making better babies. It was suggested that by throwing a few extra genes into the culture broth, a superior human being might be made, but this was always more a matter of poorly informed opinion than a practical possibility. Unfortunately, the tone of most commentaries was fearful and detracted from what ought to have been the celebration of a momentous breakthrough for science, medicine, and agriculture.

Cloning has been one of the great fictional horror stories of the twentieth century, from Aldous Huxley's futuristic society in which babies were made—from intellectual Alphas to moronic Epsilons—to Ira Levin's *The Boys from Brazil,* in which Hitler youths were created from the Führer's cells. We often make matters of high seriousness comical, perhaps to help us to cope with them. In Woody Allen's spoof *Sleeper*, a tyrant killed in a bomb blast was regenerated from a fragment of his nose. Even the naming of Dolly provided some light relief from the angst when her arrival was announced in 1997. What better way of celebrating than naming her after the singer with the most famous breasts in America?

The Scottish sheep drew a quicker and more concerted response from the world's statesmen than Saddam ever did, and journalists had a field day. *Time* ran the headline "Brave New World of cookie cutter humans" and pondered the implications of the most selfish kind of reproduction. Other newspapers ruminated about the end of sex and the redundancy of the male gender. Some even ran scare stories about cloned storm troopers marching on society.

On June 9, 1997, the White House released a press statement designed to quell fears about the misuse of biology: "Attempting to clone a human being is unacceptably dangerous to the child and morally unacceptable to our society." In New York, a representative of UNESCO commented, "Human beings must not be cloned under any circumstances." In the absence of legislation, warnings and reassurances were deemed to be more urgent—but would they be heeded? There was further uproar in the press when a retired physicist from Chicago declared that he would set up a private cloning clinic.

Not all members of the American public were as alarmist as their leaders. Many were prepared to wait and see what would happen after the froth had settled, and a few even welcomed cloning as a new reproductive liberty. According to a telephone survey for *Time* magazine, 7% of Americans said that, given the chance, they would clone themselves, and another 29% were opposed to federal regulation of cloning. Such surveys are notoriously unreliable, but it was already apparent that cloning no longer drew quite so much anxiety as it did at the beginning.

The august World Health Organization in Geneva regarded "the use of cloning for the replication of human individuals to be ethically unacceptable as it would violate some of the basic principles which govern medically assisted procreation." In Strasbourg, the Council of Europe staved off the threat of cloning by tacking a clause onto the new Code of Bioethics: "Any intervention seeking to create a human being genetically identical to another human being, whether living or dead, is prohibited." No doubt they were well-intentioned, but it is odd that institutions should condemn what nature condones every minute of the day—the frequency at which identical twins are conceived. Back in Britain there was delight that an important discovery had been made in our own backyard rather than in the United States. The authorities also enjoyed some satisfaction that there was already a British law banning cloning, although it was feared that Dolly had skipped through a legislative loophole because of the way she had been made.

Scientists had been striving to clone vertebrate animals for years. In Philadelphia in the 1950s, Robert Briggs and Thomas King had cloned frogs with a modicum of success. They replaced the chromosomes of an unfertilized egg with the nucleus from a cell of an embryo. They reasoned that if they failed at this early stage, the prospects of success with cells at more advanced stages in tadpoles and adult frogs would be zero. They did not have the advantages of modern apparatus for injecting eggs and, as so often in science, lots of perspiration was needed after the initial inspiration. Piercing eggs with a glass pipette was like spearing a soft tomato with a bean pole. Many burst, but there were occasional bull's-eyes. The best results were obtained when nuclei were transplanted from embryos or germ cells into eggs. The clones divided again and again until a completely new embryo had formed, and they soon had tadpoles swimming in the tank. But the more mature the nucleus from the embryo donor, the sooner they died. Nuclei from tadpoles or adult frogs were never successful, and the clones died at an early stage. The two scientists concluded that any attempts to clone from adult cells would be a waste of time because only germ cells and embryo cells were able to be genetically reinstructed to make a whole body again. All other cells lost their "totipotency" when the cells became more specialized in forming the skin, heart, liver, and other organs and tissues as the body matured.

A decade later, a young Oxford biologist, John Gurdon, pushed the limits of cloning a little further forward. He injected the nuclei of intestinal cells of tadpoles from another species of amphibian into eggs. Some of the reconstructed eggs developed into tadpoles, but he confirmed that the older the nucleus the more likely the clones were to die. Then he tried a more daring experiment. By recycling nuclei through embryos several times he hoped to adapt them to conditions in the egg cytoplasm so that they would be able to develop further. So he injected each egg with a nucleus and collected them after they had multiplied to reach the blastula stage. Hundreds more nuclei were generated, each of which could be injected into a fresh egg to start the cycle all over again. When clones from recycled nuclei were given full rein to develop, a few that made it to the tadpole stage metamorphosed into adult frogs that were healthy and fertile. After this exciting result, Gurdon drew the important conclusion in his paper published in *Nature* that "genetic factors required for the formation of a fertile adult frog are not lost in the course of differentiation." In theory, each of the billions of cells in our bodies could be made into another body like ourselves, but he cautiously warned that conclusions should not be extrapolated from experiments that had been tested only on intestinal cells. Likewise, we cannot be sure which *adult* cells in mammals, apart from some in the sheep's udder, can make another ewe.

Briggs and King were surprised that Gurdon's clones had gone further than their own and wondered whether he had transferred nuclei from primitive germ cells, which are easier to clone. Germ cells migrate along the tadpole's gut to their final destination in the gonad and could have been transferred by mistake. The same doubt has been voiced about the provenance of the nucleus that made Dolly the sheep: Was it from a highly specialized cell or a less differentiated stem cell? Critics have even suggested that because the ewe was pregnant, the cell might have been a rare fetal cell that had wandered into her udder, though the statistical odds of picking one up are minute. Gurdon then countered his critics by taking fully developed skin cells from a frog's hind foot since germ cells never roam so far in adult life. These clones, too, reached the swimming tadpole stage, proving that nuclei from fully differentiated cells can recapitulate development under the right conditions. They can be turned back into embryo cells and begin the process again of making descendants for all parts of the body. But the tadpoles died before the adult stage, and, until Dolly was born, it was assumed that cloning adults was a step too far.

While zoologists were cloning frogs in the laboratory, botanists were investigating the potential for cloning plants. One of the most marvelous—and sometimes shocking—aspects of biology is that a wild experiment sometimes pays off, even if we cannot say why. Engineers and physicists cannot send astronauts to the moon without thoroughly understanding gravity, but biologists have sometimes changed the course of animal and plant development with only a vague understanding of the mechanisms, and often with only low-tech methods.

The American botanist Frank Steward saved a few carrots and a coconut after shopping at the local grocery store in Ithaca, New York, and from this inauspicious start performed a startling experiment. He chopped the carrots into segments, as if preparing a stew, and dissected the conducting vessels, or phloem. Small pieces were incubated in rotating bottles containing coconut milk, which provided nutrients and hormones needed for growth. A few days later, an irregular mass, or callus, had formed and began to shed single cells that budded to form little groups of root cells. Nothing more happened unless the cells were implanted in jelly. After feeding and watering for a few weeks, they sprouted to form a whole carrot plant with a swollen orange root, just like the original.

The cells used to make new carrots are as specialized as the arteries or veins in animals. Phloem conveys sugars from the leaves to other parts of the plant and are no more a part of the seed-making flowers than our arteries are in making eggs or sperm. The biological gulf between making sap and seeds is huge, yet Steward's green fingers had managed to bridge it. He had turned the developmental clock of phloem cells back to the start, and his experiment, like Gurdon's, showed that the original genetic blueprint had not been erased. What is more, at least in plants, full adult copies can now be made routinely from single cells in the laboratory.

To understand the implications of what had been done, imagine that the genomes of plants and animals are like repertory companies. Many of the actors and actresses (genes) are used again and again in every play (cell type). But every cell has its own unique dramatis personae— genes for globin molecules are played in bone marrow cells, genes for growth hormone in pituitary cells, and so on. Sometimes the genes are switched on or off inappropriately (like a player ad libbing, or forgetting the lines), which can lead to cancer and other diseases. It is asking a lot of a company to assemble all of its original players years later for a festival to celebrate its foundation. But that is exactly what we are asking the genome to do when it is forced to behave as if it were in a fertilized egg again; we expect it to faithfully express all the right genes at the right time and never the wrong ones until the fully formed body is made. If we ever try cloning in humans we must remember that it will never be playacting, for if we make a mistake then a whole life could be spoiled.

These anxieties have often been forgotten because Dolly is the plumpest sheep on the farm and trots up to visitors like a friendly dog, nuzzling for a tidbit. She is in good condition in her third year, as is her first lamb, Bonnie, who was conceived naturally. Yet no one can predict what the future holds for either of them—except continuing affection and attention. The birth of a single healthy lamb should not make us complacent, for behind the success story there is a trail of embryos and fetuses that did not make it. Triumph with Dolly was hard won, and the experiment may be difficult to repeat. Nearly 300 eggs had to be fused with cells for the single success story. The cell of her genetic mother was 6 years old, and nobody knows whether Dolly's life expectancy will be cut short or whether she was born with a full quota of 20 ovine years. Personally, I doubt that she will age prematurely, but it is possible that the original cell had accumulated mutations. She might yet contract cancer early in life or pass on mutations that affect her descendants. Similarly, cloning may be risky in humans, the more so when the cells come from older people. But human nature being what it is, we can expect that someone will try it.

Steward's cloning experiments were always likely to be more successful than those of his opposite numbers in zoology because there are many more in of plants that have forsaken sexual reproduction than in vertebrate animals. Plant growth occurs naturally by vegetative propagation, which is a form of cloning. Not for nothing was the word "clone" coined from the Greek word for a twig. From the great aspen forests of the western United States to the buttercups in a damp English meadow, plants have successfully pursued the single life. Gardeners and horticulturists have taken advantage of the fact for centuries by making cuttings and grafts to make plants that are genetically uniform. Sexual reproduction can be bypassed, but that old philanderer Erasmus Darwin was never much in favor of chastity and celebrated plant sexuality in erotic couplets:

> Each wanton beauty, trick'd in all her grace,
> Shakes the bright dew-drops from her blushing face;
> In gay undress displays her rival charms.
> And calls her wondering lovers to her arms.

Erasmus took sex for granted, and it was hardly imagined there could be another way for the human species to increase. But the need for sex in biology is not so obvious—cloning, after all, has the advantage of reproducing at twice the rate. Where sex comes into its own is less in the sensual pleasure of the individuals concerned than in what it bequeaths to posterity. If humans choose to clone themselves, they will eventually become more vulnerable and less adaptable. That is why from almost the beginning of life on earth—from apple trees to yews and aardvarks to zebras—sex has been used in nature to create novelty and genetic refreshment. Sex forces genes into interesting new combinations, like shuffling and dealing a pack of cards. Monolineal descent abolishes variety, although this is a price sometimes worth paying when partners are scarce. When population densities are high and competition is great, hot-bloodedness tends to prevail and the sexual mode of life is the norm. But in the higher and colder latitudes, going solo can be the wiser strategy for passing on genes, and that is where more cloning species are found.

Few warm-blooded animals have adopted cloning as a normal way of reproducing themselves, and even those that have use sex to make an embryo in the first place. None have chosen to bud from an old "root stock" like the *Hydra* so beloved of biology classes in schools. But nature is so rich in variety that exceptions can generally be found. There are some animals that are not so distantly related to ourselves that have clung to vegetative roots, though never abandoning the need for eggs.

Nature's Own Clones

Armadillos look like relics of a bygone era in their horny armor but, as far as we know, no dinosaur ever reproduced its kind so strangely. Their proclivity for cloning came to the attention of two inquisitive zoologists in Texas who were puzzled about why the local nine-banded species generally produced quadruplets, and always of the same sex. They paid cowboys a bounty to catch some specimens, and in 1910, two heavyweight papers appeared in the *Journal of Morphology*. Judging by the layers of dust on the volumes in our university library, they have not been read for the better part of a century. Most of the pregnant females the zoologists examined had exactly four fetuses, each with its own umbilical cord, and in an amniotic sac. In each female only one egg had ovulated and there was a single large placenta, so there was no escaping the conclusion that one embryo had formed and split into quads soon after implantation. It is unusual to find any mammal as enthusiastic as the armadillo about cloning, but whenever it does take place in nature it always occurs through splitting. This is equally true of humans.

On the south side of Lincoln's Inn Fields, there is an imposing Georgian building that is home to the Royal College of Surgeons. On the first floor there is a museum bequeathed by the eighteenth-century surgeon John Hunter. Just inside stands the Irish Giant, Charles Byrne (or "O'Brien"), who measured 7 feet 10 inches and died at the age of 22 in 1783. He was said to have paid a large sum to an undertaker to bury him at sea in a lead coffin to avoid Hunter's clutches, but the doctor lived up to his name and somehow won his prize. Byrne has stood guard in the museum ever since, despite nearly being cremated when the German Luftwaffe scored a direct hit on the building in 1941. Beyond, Hunter's trophies stand in military rows of skeletons and pickle jars, representing the oddities of nature.

Specimen 3681 is a bottle like a glass womb full of spirits containing five little female human fetuses mounted side by side. The fetuses are alike, each with its tiny mouth agape like the subject in Edvard Munch's famous painting *The Scream*. They were born in the fourth or fifth month of pregnancy to a 21-year-old Lancashire woman in April 1786 and would have been too small to survive even in a modern neonatal intensive care unit. Dr. John White, who attended the delivery, offered the rare specimens to Hunter and mentioned in a letter that there had been only one placenta. The fetuses were presumed to have formed after a single embryo had split in the early stages of pregnancy, although we could not be sure whether there had been only one ovulation or if they were identical quintuplets.

The genetic identity of ancient specimens can now be resolved using DNA fingerprinting, although, after 200 years, molecules can be in advanced stages of decay. If you wanted to leave your genome to posterity, it was better to be embalmed in Luxor than pickled in London, though deep-freezing is now best. When the five fetuses were analyzed by DNA fingerprinting a year ago, biologists had expected to find that they were identical, but in the event only two were. There must have been four ovulations and four embryos, one of which later split to make identical twins.

Although the Hunterian quintuplets have turned out to be less interesting than expected, five other remarkable sisters were born in Canada, and three are still alive. On October 7, 1933, the Dionne family made an emergency telephone call to Dr. Dafoe's surgery in Callander, Ontario, urging him to attend a home delivery. As he put down the receiver he had no idea that a sensational birth was in progress, nor did he sense anything out of the ordinary when he crossed the threshold of the Dionne home. "I arrived to find the home in confusion, no preparation made for confinement, except the tea-kettle boiling on the stove. Two babies had already been born, and a third was just making its appearance over the perineum. Two neighbors were acting as midwives. The father had disappeared." In the next few minutes two more babies were born within their own sacs. Finally, a single, large placenta emerged. Amazingly, all five babies were healthy. These are the only such set of quintuplets ever known to have survived, and tests for blood groups and DNA fingerprints revealed all five sisters to be identical.

The girls were raised in a hospital that was built on the instructions of Dr. Dafoe across the street from their family home. Visits from their parents, brothers, and sisters were strictly rationed. The little girls all had beautiful brown eyes and black ringlets, and the staff dressed them the same and trained them together. Thousands of members of the public seeking an entertaining diversion from the hardships of the Depression descended on "Quinland" during their vacations and weekends. The girls quickly became celebrities and were in demand by advertisers eager to use their famous names and images on brand products. Revenues topped $200,000 a year at the peak of their popularity. The three surviving sisters still live together in Montreal and recently filed a successful suit against the Province of Ontario for childhood exploitation as medical curiosities. If this is the fate of natural clones, how much worse will the first artificially cloned children fare?

Clone may seem too pejorative an expression to use of human babies, but that is exactly what genetically identical siblings are. Their existence should inform and calm the debate about unnatural cloning, for twins are not rare. Most twins, however, are produced by two ovulations, in either the same or opposite ovaries. When eggs are formed and fertilized separately, they are as genetically dissimilar as brothers and sisters from singleton pregnancies and as likely to be of opposite as of same sex. Identical twins have matching sets of genes both in their nucleus and in their mitochondria, unlike artificial sheep clones, which inherited their mitochondria from the egg's cytoplasm.

Identical twins usually arrive out of the blue. The splitting of one embryo into two is a chance event that can occur in any woman of any race and at any age. Though still relatively unusual, it happens more often when a couple is undergoing treatment for infertility. Two embryos transferred to the womb can occasionally make three babies, by one splitting into two. Or three can make four. It is thought that the split occurs when the embryo is escaping from its shell-like membrane, which may have cracked during freeze-storage or as the result of an embryologist cracking a hardened shell to help a trapped embryo. Neither the doctor nor the patient should be held to blame for the whims of nature, but woe betide the one who deliberately sets out to split an embryo.

Human Hatcheries

To judge from newspaper reports, the first human "cloning" experiments were already underway at George Washington

University five years ago. Far from aiming to make a set of identical children, the scientists at the clinic hoped to improve treatment for infertile couples wanting to have a single child by IVF. They wondered if by splitting embryos at the four- or eight-cell stage and wrapping each one in its own artificial membrane to make miniembryos, more pregnancy attempts could be made for lower cost and effort. The cells in early embryos are still totipotent, and each can, in theory, make a normal baby. Mindful of public sensitivity, the research team used only embryos that had been fatally flawed by fertilization with more than one sperm. But the team still became the subject of controversy when the story broke in the newspapers. We now know that it is a waste of time to break precious embryos into fragments for it is unlikely to increase pregnancy rates, and worries about making identical quadruplets or octuplets were completely unfounded.

However futile in practice, these experiments primed the pump for the debate when Dolly arrived a few years later. For the world of biology, 1997 was A.D. 1, year 1 After Dolly, as it marked the beginning of a new era of research—and uncertainty. Was sex really necessary to procreate? What would have seemed an absurd question previously was now being discussed in conferences and coffee shops around the world. The processes that Walt Whitman had assumed to be so primitive and necessary now seemed old-fashioned:

> *Always the procreant urge of the world.*
> *Out of the dimness opposite equals advance,*
> *Always substance and increase, always sex . . .*

The normally cautious pioneer of sheep cloning, Ian Wilmut, has said that human cloning "could happen if we wish it to happen." It is another matter entirely whether it should happen. Like most scientists, Wilmut is implacably opposed to it, and I can see few circumstances, if any, in which it would be helpful or justifiable on medical or compassionate grounds to clone an adult human being. If our goal is to make healthy babies and to help infertile couples, then we should reserve precious eggs for those who need them most. Nevertheless, it is worth considering some of the suggested ends—many of which are far-fetched—to which cloning might be applied.

There is the suggestion that since kidneys and other organs for transplantation are in short supply, a fetus could be cloned to save lives. An organ from a personal clone, being identical in the all-important genes involved in transplant rejection, would not require suppression of the immune system because the tissue match would be as perfect as an identical twin's. But there would be an insurmountable delay in waiting until the organ was mature enough to be used for a kidney transplant in a child. This problem is admittedly less where bone marrow is required for a severely anemic child. The cynic would say that if the patient died, there could be the compensation of the best possible replacement technology can provide. It has even been suggested that clones could be made with the express purpose of providing transplant organs, perhaps after eliminating certain genes to make headless monsters. This would be an interesting dilemma for ethical philosophers, as it would seem cruel to deny people the use of their own cells for spare-part surgery. Another idea is that brain cells could be "harvested" from cloned fetuses for transplantation to people with Parkinson's disease.

But cloning might also have an application as a fertility treatment. It has often been pointed out that cloning could help a woman who has no ovaries of her own or has had a premature menopause to conceive. But would it be any advance on egg donation? Instead of her baby inheriting some of her male partner's genes, it would have only her own, and the health risks for the child are likely to be much higher than with conventional egg donation. Such genetic exclusivity might of course be an attraction for some people who want to avoid the male contribution altogether.

Whatever the motives, cloning will have serious implications, causing our fundamental sense of kinship to go awry.

Cloning is even more genealogically perplexing than egg and sperm donation and surrogacy: lineages shift horizontally rather than flowing vertically from generation to generation. Luckily, Dolly does not have enough grey matter to be confused. Is she the daughter of the ewe that bore her, of the one that donated the egg, or of the one that posthumously donated the cell nucleus? And since there was no sperm at conception, was she born fatherless? Or was her father the ram whose genes she inherited in the nucleus from an earlier generation, or the scientist who directed the experiment? If the conundrums of sheep genealogy send our heads into a spin, the bizarre implications of human cloning will certainly cause even more confusion. Whither the traditional nuclear family of Mom, Pop, and the kids? Cloning would render the notion of "children" meaningless, giving us copies of existing people instead, but possibly decades after the cells of those people had been frozen. Natural cloning in the shape of genetically identical twins is taken for granted and does not disturb the genealogical equilibrium, but laboratory cloning raises issues that have never been encountered before.

But within strict guidelines, I believe that there is a role for laboratory cloning—not to make copies of existing people, but ironically to assist the procreation of unique beings. If a couple has but a slim chance of producing a desperately wanted child and only a few precious embryos are available, they might use cloning to increase their prospects. Although we have seen that splitting embryos and growing the products on their own is fruitless, dividing them into four for fusing with enucleated donor eggs might improve their chances of producing a fetus. Such is the natural desire for children that childless couples who can afford this solution are bound to seek it. It may be a back door to cloning, but it is hard to find a clear justification for denying them such treatment.

This occasional use of donor eggs for improving fertility will not place a heavy demand on scarce supplies, but the egg shortage will, for the foreseeable future, limit any ambitions to extend cloning. These are the rarest cells in the body and will always set a limit on cloning while they are needed. Only one egg is ovulated per month, and then only for the 35 years between puberty and menopause. Even a lifetime's production of up to 500 eggs would make the equivalent of only one Dolly at current success rates.

Aldous Huxley alluded to these limitations in *Brave New World*. In his imaginary society the authorities forbade natural fertility and women were required to sacrifice their ovaries at puberty so that eggs could be ripened and cloned in vitro. The director of the Central Hatchery and Conditioning Centre in London summarized the benefits to a group of green students: "In nature, it takes 30 years for 200 eggs to reach maturity. . . . But Podsnap's Technique has immensely accelerated the process of ripening. They could make sure of at least 150 mature eggs within two years. Fertilize and bokanovskify—in other words, multiply by 72—and you get an average of nearly 11,000 brothers and sisters in 150 batches of identical twins, all within two years of the same age." Many years later, the pioneers of IVF tried to increase the egg supply for their patients using a "Podsnap" method. They collected immature eggs and cultured them in a nutrient fluid laced with hormones, but sadly this did not work quite as well as it had in fiction. They soon abandoned the attempt in favor of collecting eggs from the ovary after treating patients with injected hormones.

The hormonal stimulation of young women is currently the only source of eggs for cloning, though in Britain, where payment of donors is forbidden, eggs are already very scarce. In the United States, women are paid $3,000 to $6,000 for a clutch of about ten eggs, and consequently an egg market has sprung up to meet the demands. Even at the cheaper rate and with professional fees included, the bill for collecting the 300 eggs needed to make a clone like Dolly would be over $100,000—and we can only guess how many eggs would be needed for human cloning. When the costs of the other procedures are included—cell freezing, egg fusion, and embryo culture—there would not be much change left from $250,000. Cloning will be the privilege only of the rich, unless eggs are found more abundantly and cheaply.

In our laboratory, we are trying to ripen human eggs in vitro. The aim is not to clone but to provide more eggs to donate to infertile or cancer patients who freeze-banked their ovaries before having sterilizing chemotherapy. Our dream is to ripen some of the thousands of eggs in a young woman's ovaries, most of which normally go to waste. Even a small biopsy contains hundreds of tiny eggs, which can be stored until required, and then released into a culture fluid for growth to maturity.

Egg farming is not a cloning technique because it increases the number of genetically unique cells available, but increasing the supply of eggs could make cloning a practical technique at some time in the future. Until then human cloning will have little impact on society.

Unity or Diversity?

Every minute of our lives, millions of cells in every organ in our bodies are busily engaged in making an exact gene copy before they divide. But making faithful carbon copies of cells in which the number and location of every molecule is identical is another matter. The DNA is simple compared with the baffling and marvelous complexity that exists downstream in the cytoplasm of cells. From the DNA template in the nucleus, tens of thousands of different RNA molecules are transcribed as messengers to carry instructions for making proteins in the cytoplasm. The process is far more complex than a cellular mailing and translation service. If there were ever a technology sophisticated enough to make the analysis, we would find that every cell in our body has something marking it out as an individual, even though all contain the same genes. If difference is the rule at the cellular level, how much greater must it be between whole bodies?

It is self-evident, of course, that genetically identical twins are more alike than are other siblings, but the degree of likeness is often exaggerated. Twins are dressed alike, taught the same manners, and encouraged to do almost everything together, reinforcing the illusion of sameness. Beneath these social embellishments, however, are two people whose degree of similitude is striking only when compared with the enormous variety of human form and behavior.

Few people have the opportunity to get to know many twins before external influences have left their imprint. The director of the Multiple Births Foundation in London, Elizabeth Bryan, is one, and so convinced is she about the uniqueness of individuals that she urges us to abandon the word "identical" for any twins. Babies that have formed from splitting embryos, she argues, should be called "monozygotic" or MZ twins, and those derived from separate embryos, "dizygotic" or DZ twins. The unique identity that emerges despite a shared genetic inheritance should allay our fears that cloning would produce exact replicas of people. Yet Bryan worries about the attitudes of the families of "identical" infants and society generally: "Despite increasing awareness of the importance of twins being encouraged to develop their own individuality from the start, many parents find it difficult to put it into practice in the first year."

Twins pay a price for our fascination. The IQ of twins is no less than in other children, but they learn to speak more slowly because they often use a secret language that excludes others and reinforces the knot between them. This close bond can mean trauma when twins are separated through marriage or death in later life. In one case, a surviving twin had to pack away all family photographs because the twins had never been apart, and the image of his brother reminded him of the pain of separation. A 22-year-old man could never shave in front of the mirror again, as he saw his dead brother staring back. Twinship carries tension as well as fraternal love. One set of MZ twins excelled equally at sport, but only one could be the champion at school, and this led to a love–hate relationship. In *Through the Looking-Glass,* Alice met twins who were not always the best of pals, even though they were always in each other's company:

Tweedledum and Tweedledee
Agreed to have a battle;
For Tweedledum said Tweedledee
Had spoiled his nice new rattle.

Contrary to popular imagination, MZ twins can sometimes be quite dissimilar in size and appearance at birth and may diverge even more later on. If notable differences exist between babies from the same pregnancy, how much greater would the variations be in clones that are made 10, 40, or even 70 years apart and gestated in different wombs? The three surviving Dionne quintuplets, Cecile, Annette, and Yvonne, have an obvious family resemblance but otherwise do not look strikingly similar nowadays. There are stories about rare MZ twins who were separated when they were young but still chose the same life style, had the same tastes in food and drink, and pursued the same occupation. But exceptions do not prove a rule, and many stories that fail to fit the mold go without remark.

Although MZ twins are always formed by splitting an embryo, most develop in their own amniotic sacs and in consequence have slightly different experiences of pregnancy, which might leave impressions later in life. Physical intimacy is most intense in the 3% of MZ twins who occupy the same sac, and some of these even share parts of the same body. Nature and nurture were never closer than in Siamese twins, although even they can be remarkably different.

The original "Siamese twins," Chang and Eng, were Chinese, though born in Bangkok in 1811. They were rescued from execution when they were 13, and an English merchant sent them to England where they became celebrities and were introduced at court. They fell in love with English sisters and were not the last such duo to have problems with priests who refused to marry them on the grounds that it would be bigamy. But the double union was consummated and they produced 22 children between them. They had a flair for entertainment and found fame and fortune in the United States, where they toured with a circus, astonishing audiences with their head-over-heels acrobatics. Although they were attached at the chest and upper abdomen, they lacked a strong physical resemblance and possessed very different temperaments. Chang liked to drink whereas Eng was a teetotaler; it was fortunate that they were so forbearing of each other. One English gentleman who met them was so impressed by their differences that he celebrated them in verse:

> *Now Chang was slow, he learnt his letters*
> *As if his memory moved in fetters . . .*
> *But Ching [Eng] was hasty, quick and clever,*
> *His soul's glad stream flowed out for ever.*
> *Chang, tho' austere, was mild in bearing,*
> *Calm as a smile from Lady Bury;*
> *But Ching perpetually was swearing,*
> *And fidgeting himself to fury.*

> —Sir Edward Bulwer-Lytton, *The Siamese
> Twins—A Tale of the Times* (1831)

Other examples confirm how unalike Siamese twins can. The Filipino twins Simplicio and Lucio Godina (1908–1936), like Chang and Eng, enjoyed show business success in America, where they danced and performed together on roller skates. But Simplicio was heavier, with a broader head and face, and was less lively than his brother. Then there were the contemporary Hilton sisters. One was 2 inches shorter and considered prettier than the other. There was also a marked difference between their scores on intelligence tests.

Mark Twain reflected on the paradox of diversity in unity in *Puddin'head Wilson and Those Extraordinary Twins.*

His fictitious twins, Angelo and Luigi, had one body but two personalities, like the original Siamese twins. They shared a pair of legs and helped to feed each other at mealtimes. But one was teetotal and became inebriated when his brother drank, one liked to stay up at night reading Tom Paine's *Age of Reason*, and each liked to sing a different song at the same time, usually discordantly. Finally, after the trials of trying to be different, they shared the same fate on the gallows when one was hung for his crimes.

As a species that normally produces babies one by one, we are fascinated by twinning and higher multiple pregnancies, especially when the offspring are genetically identical. Individuality—in body, mind, and soul—is something we normally take for granted, and we feel uneasy when faced by genetic sameness. But twins—even Siamese twins—have as strong a sense of identity as the rest of us, despite well-meaning but potentially harmful pressures from others to deny it. MZ twinning and cloning are also flash points in the "nature or nurture" debate. Arguments about the heritability of intelligence still rage and genetic pessimism stalks society despite campaigns by well-known scientists to eradicate an oversimplified distinction. Genes and environment are like Siamese twins—locked together in a dialogue.

But a third factor affects the way we are, and it is often forgotten—the capricious harlequinade of nature. Even if we were able to control both the genes and the environment, the random variation factor that occurs during development limits our ability to make identical copies of living things. Mathematicians call this factor "structural variation" because it is inbuilt. It is little understood by biologists because we cannot manipulate contingent nature and have to accept an irreducible amount of unpredictability. Some of this variation is subtle and goes unnoticed. For example, on the edge of an armadillo's armor plates are grooves that show different patterns, even among "identical" quads. Cloned cattle have patches of pigmentation in their hides that are unique to each individual. And not all variations are "cosmetic." The numbers of eggs in the ovaries of genetically identical animals, and probably humans too can vary severalfold at the same age. This affects their fertility and the time of menopause.

The sheer complexity of the human brain demonstrates more than anything else the impossibility of making an identical product. There are approximately 100 billion neurons, and each is connected to about 10,000 others, providing zillions of pathways for electrical impulses to travel. These connections cannot all be mapped by the 80 million genes in the genome, so it is absurd to imagine that the brains of MZ twins—or clones—are identical. Indeed, although most pairs of twins are right-handed, some have a right- and a left-handed member; the Dionnes differed in this respect, too. Such a complex organ leaves vast scope for variations in behavior, cognitive ability, and motor control.

Biology abhors uniformity as much as physics hates a vacuum. We are all fully unique beings whether we are genetic originals or otherwise and even if we share part of a body with a twin. There can never be a "perfect clone." If we could start our lives over again from the moment of fertilization, perhaps the only thing we could be sure of is that we would turn out differently. This is reminiscent of an illustration that Steve Gould used for the contingent nature of the universe. In *Wonderful Life,* he imagines rewinding the tape of biological history and watching the story of evolution again and again and muses that the amazing explosion of life forms and extinctions would happen differently every time.

Far from deriding our biology for its endless variations and harmless "imperfections," we ought to celebrate the protean nature of life. Perhaps we already do and this is why the notion of cloning, of all the technologies mentioned in this book, is the most abhorrent. My own reservations about a technology that has its parallel in nature may seem incongruous, given my commitment to such an "unnatural" technology as assisted reproduction and my hopes for germ line therapy. But I believe the primary goal of reproductive medicine and biology is to help people bear a child with the best possible start in life—physically and psychologically—and to bring diversity to family life. We have yet to see any evidence that cloning can serve those purposes. Indeed, the evidence is much to the contrary, and it is difficult to imagine in what ways a child's interests would be served by this technology.

6

Sex Selection

Sons and Daughters

A more or less fixed ratio of one male birth for every female in the population makes biological sense. If a mutant that favored, say, male offspring arose in nature there might be an initial advantage in winning more mates. But as females became scarcer they would be at a premium, and the advantage of the mutation for males would decline and the sex ratio would edge back toward balance. Yet behind this simple logic there are exceptions that still defy explanation.

The big red female eclectus parrot (the male is green) of New Guinea and northern Queensland can produce long runs of one sex or the other. One female kept in a zoo produced 20 sons in a row and then 13 daughters, a pattern very unlikely to have arisen purely by chance. The unstable sex ratio of alligators was also a mystery until it was found that the temperature of the nest affected the development of the embryonic gonads, and males like it hot! In humans the ratio at birth is close to parity—about 105 boys for every 100 girls born. Since fewer male fetuses survive than female ones, there must have been an even greater imbalance at conception, but we have little idea why. The statistics of large families in historical populations confirm that the birth of a boy or girl does not strictly tally with the binomial theorem, which predicts random events like heads and tails after tossing a coin. Such evidence hints that with the sex ratio we are not dealing strictly with chance events. Likewise, after the great slaughter in the two world wars, there was a change to a slightly greater preponderance of males born in England. In Europe today a drift is occurring, but it is in the opposite direction.

There are suspicions that hormones and paternal age are responsible for such variations. After World War I, fewer young men returned home, and during World War II, couples

commonly put off having a family. So the postwar generations were born to older fathers. This conforms with a
recent study at Liverpool University that has shown that
more boys are born to families in which the father is 5 to
15 years older than his wife. The ratio reverses with an
older wife. This could be an explanation for the tendency
of monarchs, tribal leaders, and the Hebrew Patriarchs to
produce male heirs: all tended to take brides much younger
than themselves. The English aristocracy seem to have followed the lead. The 1901 census observed: "If the population of London be divided into three portions exhibiting
graduated poverty, it is found that the proportion of male
to female infants produced is lowest in the poorest portion,
highest in the wealthiest portion, and intermediate in the
intermediate portion. The proportion of males is highest of
all in a number of births taken from *Burke's Peerage,* where
the nutrition may be supposed to be of the best." Their
lordships may not have known how they managed it, but it
suited them to keep the power in male hands. By contrast,
in poor families it is better to produce girls, who stand
more chance of marrying someone of higher social rank, so
improving the fortunes of their relatives.

For a long time it was not understood which partner
was responsible for making the child a boy or girl. At the
end of the nineteenth century the Cambridge pioneer of
reproductive science, Walter Heape, thought that sex genes
were biologically emancipated and that eggs and sperm
came in male and female forms. This news would have
pleased his contemporary, Emmeline Pankhurst, and other
suffragettes, but soon after the turn of the century, the sex
chromosomes were discovered, setting the foundations of
our understanding of how sex is determined.

The sex of a baby is fixed according to whether a male
or female sperm fertilizes the egg. Half the sperm produced in each testis carry a Y chromosome, which is a
rather short and mainly degenerate stretch of DNA carrying the genes needed to make the testes and produce

sperm. The other half carry an X chromosome, like the eggs. If sperm and egg meet by chance there should be an equal number of XY, or male, and XX, or female, embryos. The type of sperm therefore fixes the sex of the embryo, although the ratio of males to females at birth will, of course, change if one sex or the other suffers a higher pre-natal mortality.

Until these facts were known, attempts to control the sex of offspring were left to imagination and invention. For 2,000 years, the most respected medical authority was the Greek physician Hippocrates, who believed that the right testicle produced sons. He advised that if a daughter was wanted the man "should tie off the right testicle as much as he can bear." He also recommended that "if a man wants to sire a son he should have relations with his wife at the end of her period and should thrust as hard as he can until ejac-ulation." No one knows how many earnest fathers down the ages have tried these methods, but probably about half of them had achieved the outcome they had been hoping for.

There have been plenty of other recipes for making boys—high potassium and sodium diets, vaginal douches containing vinegar, turning the nuptial bed to face the north wind—but no one has yet hit on a reliable one. Strength and excess supposedly characterized the condi-tions favoring male conception. More recently there have been claims that the male sperm has a more positive elec-trical charge. How understandable, then, that feminists should take a jaundiced view of all this nonscience.

According to another theory, male sperm swim faster but die younger than their sisters (like human siblings), so by judiciously choosing the time of intercourse during a woman's menstrual cycle, a couple can alter the odds of having a boy or a girl. If insemination occurs around the middle of the cycle, when a ripe egg is awaiting the arrival of the first sperm, they are more likely to have a boy. If insemination occurs earlier, then the odds will favor a girl, because female sperm will be surviving in larger numbers by the time the egg pops out at ovulation. Like the other theories, this one is better in the telling than in the test. If differences exist between the sperm in their swimming ability or longevity, no laboratory test has yet detected them. In any case, the method may risk the health of the embryo if a couple is trying to make a baby girl. It is much safer to aim to synchronize insemination with ovulation so that both egg and sperm are as fresh as possible when they make an embryo.

After determining the sex there comes the desire to predict it. The ancient Egyptians had a method for testing for both pregnancy and sex at the same time. The woman sprinkled her urine on a few grains of barley and emmer (a kind of wheat). If both sprouted, the test was positive for pregnancy; if only the barley did, it was a boy; if only the emmer, then it was a girl. The method was still being used in nineteenth-century Germany and, when tested under laboratory conditions in the 1960s, was found to be 70% accurate for confirming pregnancy but totally unreliable for predicting gender.

There is usually plenty of domestic speculation about the sex of a child. When my wife's pregnancy showed in front rather than more evenly all around, we were told that we would have a son (as we did). I have heard other people say that if a ring is suspended over the woman's tummy and it moves back and forth like a pendulum, she will have a boy, and if it moves in a circle, a girl. Like the fanciful ideas for controlling the sex of a child, these tests owe more to our gender assumptions than to science.

Superstition and bias have not yet been swept aside, but technology now provides parents with a prediction—though not control—of their baby's sex that is virtually 100% reliable. Amniocentesis and chorionic villus sampling (CVS) can be used to diagnose the fetus's chromosomal sex, and ultrasound scans allow parents to see whether their baby has a penis, even before pregnancy has reached the halfway stage. This is information they can now count on and use as they may.

Striking a Balance

Whether a mother delivers her baby in a hospital, in her own bedroom, or in the bush, she wants to know immediately whether it is a boy or a girl. Likewise, this is the first or the second news flash that the midwife transmits to the anxious father, who passes it on to grandparents, aunts, and uncles. After assurance that the baby's health is good and before starting to compare family resemblances, the sex matters most of all.

Being a male or female may not be the most important fact about a baby, but it is the most obvious. Rarely is there any ambiguity about the sex at birth, because the testes are stimulated before birth by hormones and are larger than at any time afterward until puberty. There are no equivalent anatomical tokens to tell us whether babies will turn out to be clever or stupid, jovial or serious, sociable or introverted, and if there were, how differently we would treat him or her. Whichever sex is assigned, the label is likely to stick for life, and the few who try to change find the journey harrowing. From their very first days babies in our hospital are dressed in pink or blue, as if to instruct everyone how they should be treated henceforth.

There are fewer doubts when designating sex than any other feature because there are only two alternatives—although that has become a point of controversy. In an age when celibacy was supposedly more common, the Reverend Sydney Smith once jested: "As the French say, there are three sexes—men, women and clergymen!" So fundamental in the human psyche is *la différence* that some European languages have evolved a sexual dichotomy for different kinds of pronouns and nouns, and the word "gender" has taken on new meaning. According to Steven Pinker, author of *The Language Instinct,* the word was originally used linguistically for "kind . . . (and) has been pressed into service by nonlinguists as a convenient label for sexual dimorphism; the more accurate term sex seems now to be reserved as the polite way to refer to copulation." That may be true of the majority, but biologists have stubbornly chosen to work *sex* harder and apply it to the differences between males and females even when polite conversation now recommends *gender*.

It is difficult to express any opinions on this subject without drawing criticism for being sexist, and this is an occupational hazard for reproductive biologists. *La dif-férence* begins at conception, and its biological roots sprout almost everywhere in our nature and culture. It is not only our biology that is colored by sex; the way we talk and behave is affected by it too. It is easy to find a bias in the way language is used. Even the French word for sex comes only in a masculine form. It is still commonplace to say *le sexe faible, le deuxième sexe*, or *le beau sexe*, with the implicit discrimination these phrases carry.

The preference for one sex—mostly male—has been and still is so prevalent in large tracts of the world that it colors all rational debate. India is said to be 40 million women short of the numbers that a normal sex ratio would assure. This deficit—almost equivalent to the total population of England and Wales—is caused by the abortion of female fetuses, infanticide, and the early death of girls who are heartlessly neglected or abandoned. Misogynist attitudes are as widespread among women as men, and midwives are often responsible for dispatching unwanted baby girls. Where a woman's lot in life is hard and poor families can be ruined by the burden of dowry payments, there is complicity between wives and husbands in dreadful acts that they believe are in the best interests of everyone, including the victim. Similar attitudes are common throughout Southeast Asia, especially in rural areas, and millions of Chinese girls have been killed in spite of all government attempts to intervene.

A preference for males has a long history. In India this attitude was firmly embedded by the time of the British Raj and was at least as prevalent among the higher castes as in the primitive tribes. According to a nineteenth-century visitor to Benares, "every female infant born in the Rajah's family of a lawful wife, or Rani, was drowned as soon as it was born, in a hole in the earth filled with milk." The Ranas were said to have no grown-up daughters for more than a century. In the local Rajpoots tribe, female offspring were invariably killed "except in rare instances when no other issue existed." Across the country in Bombay, Governor Mountstuart Elphinstone summed up the problem rather pompously: "We must . . . be content to follow the footsteps of our predecessors in their most meritorious endeavors to discountenance this enormity, and we may flatter ourselves that, as the manners of the people become softened by the continuance of tranquillity and good order, they will gradually discontinue a practice which is not more inconsistent with reason than repugnant to natural instinct." The official response to the ancient Hindu practice of *suttee* (the self-immolation of widows on their husbands' funeral pyres) was more firm: the British banned it in 1829, though such deeply rooted customs are not extinguished easily and cases are still reported.

The government of India continues the battle to save female lives through a combination of stick and carrot. Poor families earning less than 9,000 rupees per year ($210) are now offered 500 rupees for every daughter and given incentives to send them to school to reduce the huge literacy gap between boys and girls. Yet despite these efforts and the passing of the Dowry Prohibition Act (1961), daughters are still regarded as financial liabilities or as "another family's wealth." Businesses offering ultrasound scanning for fetal sex testing are openly advertised and often situated close to abortion clinics. It remains to be seen whether the Pre-Natal Diagnostic Techniques Act (1996) will be any more effective as a deterrent or will merely drive practices underground.

When people emigrate to another country, they take with them many of their cultural values, and this can present a dilemma for the authorities. Leading Asian women in Canada have tried to bring about a more welcoming attitude to daughters in their communities. A play, staged in

Vancouver and Toronto in the Punjabi language, depicts a husband and wife who desperately want a son to carry on the family name. The man threatens to leave his wife and their 10-year-old daughter unless she aborts the female fetus she is carrying to try again for a son. As the drama unfolds, it is clear that there is a campaign message. The lead actress, Darshan Mann, explained, "I either have to go for an abortion or risk my marriage . . . it's very emotional."

In countries with pregnancy screening and liberal abortion laws, doctors can be hoodwinked into providing a sex selection service, and this probably happens more often in Western countries than we imagine. This is a dilemma for the medical profession. Doctors are required to offer pregnancy screening and abortion services and cannot deny patients information in which they have a personal stake, even though it could easily be misused. Information gained about the baby's sex in one center can be acted on surreptitiously at another.

Eliminating unwanted offspring has hitherto been the easiest way to select a child's sex. But reproductive technology promises another, more acceptable means of choosing a baby's sex—sperm sorting before conception. Would this tip the bias toward males even further and produce a disastrous population imbalance? Most people who have large families want a mixture of sons and daughters, and most patients attending gender selection clinics in the West, generally of Asian extraction, already have children of the same sex and seek another of the opposite sex. Once the woman has borne a male child, the couple is less concerned about and more indifferent to the sex of any later siblings. Surveys of Europeans and Americans suggest that the population impact would be negligible. The Dutch Health Council concluded that sperm selection would lead to more mixed families but few people would actually avail themselves of it: "It would seem improbable that the position of women in our society would noticeably worsen as the result of the availability of a technique of sex-selective insemination." In a British survey women were asked during the early stages of pregnancy: "Do you mind what sex your baby is?" To which a majority of 58% replied, "No." Among those who expressed a strong preference, 6% wanted a boy and an equal percentage wanted a girl; there was only a hint of male bias in even the minority sample of Afro-Caribbean and Asian women. Perhaps the most surprising result was that over half did not even want to know the sex of their child before birth, especially those who were expecting their first child.

Historically there have been a number of exceptions to male dominance. Until the arrival of the Greeks, women in ancient Egypt had access to all the professions, and the wife of a pharaoh might occasionally take power after the death of her husband. Queen Hatshepsut of the eighteenth dynasty oversaw the expansion of her empire during a long reign, just as Queen Victoria did 3,000 years later in England. More primitive peoples might be expected to be more male-oriented, but the social anthropologist Claude Lévi-Strauss, drawing on evidence from tribes in the Trobriand Islands, concluded that the value attached to each sex in society is highly complex: "They take pride in numbering in their clans more women than men, and are unhappy when there are fewer women. At the same time, they hold male superiority as an accepted fact."

As we saw in Chapter 3, people have manipulated the sex ratio in their own interests to make warriors or farmhands or to earn income from dowries. The cultural impact of the natural sex ratio is complex and changeable. In any society there must come a point at which any marked male bias will lead to a surfeit of bachelors and a corresponding increase in the value of female children. What is more, the balance of social and economic advantages of having sons may change. This already seems to be happening in our society, where the supposed female virtues of diligence, networking, and cooperation are stressed.

While we cannot be sanguine about attitudes everywhere or the methods used, interference with people's wish to choose a child's sex at *conception* ought to be reserved for when there are strong reasons. The benefits might even outweigh the costs in some traditional societies, since the opportunity to choose might encourage them to stop having babies earlier rather than to keep trying in the hope of a boy and producing a larger family than they desire and more mouths to feed than they can afford. "Family balancing" is the more acceptable face of sex selection in the West. Nature has dictated the order in which boys and girls appear in a family, but it is not obvious whether any harm would be done to the children or to society if choice was possible. It seems hard to criticize a couple who, after a string of children of one sex, want one of the opposite.

The Dream Screen

The obstetrician–gynecologist who knows most about identifying the sex of fetuses is director of Koala Laboratories, based in an unprepossessing strip-mall building in Blaine, Washington as well as in California. John Stephens is an expert ultrasonographer and claims to be able to distinguish male and female fetuses at an earlier stage of pregnancy than anyone else. In 1990 he filed a U.S. patent for fetal anatomic sex assignment, a technique "based on pattern recognition that allows identification of external genitalia during the gestational age range 12–14 weeks" (dated from the mother's last menstrual period). Many others in his profession were skeptical about a method that could identify the sex of a fetus a full four weeks earlier than in conventional practice. Independent trials are needed to substantiate his claims, which have attracted unsavory press attention and kept him at arm's length from colleagues in the medical establishment. Theoretically, it is possible to identify fetal sex even earlier than 12 to 14 weeks because the testes begin to differentiate a month before, but it is doubtful whether this can be achieved by scanning, and Stephens is working at the limits of the technology.

Through scanning thousands of pregnancies, perhaps he has acquired the sort of uncanny sense of perception that some bird watchers have for identifying a rare species with a fleeting glance. We shall shortly see that a more scientific approach may eventually advance the time of sex recognition from the present limitations.

Stephens' patients are mainly Sikhs from Vancouver, Toronto, and California, and they come because they are eager to know whether the mother is expecting a boy or a girl. He strenuously argues that his clinical service is for fetal sex determination and not sex selection, but he works on the edge of an ethical dilemma: Canadian law forbids the use of abortion for nonmedical reasons. Nobody knows how many patients return home glumly and then decide to request an abortion from their local doctor on some other pretext. Stephens defends his work by declaring that he is in favor of enhancing patient autonomy in decision making. He is correct in recognizing a trend toward more self-determination and less paternalism, but his emphasis is not widely welcomed in his profession.

Some doctors in Britain are concerned about the danger of declaring the sex of the baby so early and prefer to reveal it only in the later stages of pregnancy. A child's sex is not normally a feature of pressing medical significance. Only in rare cases where there is a history of disease linked to males in the family is action required. This might be for a serious condition such as Duchenne's muscular dystrophy or hemophilia. But do doctors any longer have the right to deny a patient information about her own pregnancy? Or should judgment even be passed on the wish to have a child of a specified sex? The ethics committee of the American Society for Reproductive Medicine found "highly problematic the use of gender selection to achieve 'family balancing' or other preferential goals based on nondisease traits. However, it may be premature to declare that there are absolutely no circumstances under which gender selection should be used." Evidently the jury is still out.

The debate is swept along by the wind of technological progress. It is already possible to diagnose fetal sex after 10 weeks of pregnancy using CVS, several weeks before amniocentesis is feasible. But the technique carries a slight risk of harming the fetus and should only be used for diagnosing a disease rather than the comparatively trivial purpose of determining sex. But inevitably there are cases where it has been used for social rather than medical reasons. An Israeli woman who already had two sons underwent CVS for testing fetal sex when she became pregnant with twins while they were working in the United States. The twins, a boy and a girl, had separate placentas. The boy was terminated because the parents feared that a son would be more likely to be killed after conscription into military service. This is not a persuasive reason for making such a serious decision. A lesbian couple arranged conception using donor sperm that had been subjected to a technique that was claimed to enrich the fraction of sperm with the female-determining, or X, chromosome. They, too, were expecting twins, and CVS diagnosed a boy and a girl. They approached another clinic and asked an unwitting physician to selectively reduce "fetus B," which they had thought was the male, at 11 weeks. After the surviving fetus was checked, they discovered that it was the "wrong" sex. So they aborted that one, too, and decided to start again. These stories are regrettable in themselves and also discredit some genuine arguments in favor of family balancing.

In the future it may not be necessary to use either the invasive CVS technique or the gentle method of scanning since fetal cells entering the mother's circulation can be used for sexing the baby at the same time as for genetic testing. A simple blood sample from her arm as early as 5 to 6 weeks of pregnancy contains a few extremely rare fetal cells that can be analyzed for unique DNA sequences on the Y chromosome to check whether a male is present in the womb. It is more tricky if there are twins. When one of our staff and his wife were expecting a baby they decided to test the technique out of scientific curiosity. As expert molecular biologists, they developed the highly sensitive method of DNA amplification to test for the Y chromosome. But the technique was blighted by false positive results in the control tubes that contained "pure" water, so the results from blood were untrustworthy. Try as they might, the false signal was never eliminated, and we concluded that our lab was "contaminated" with fragments of male DNA that were presumably being shed in skin cells by staff. Short of recruiting nuns from a closed religious order to perform the laboratory tests, this screening seems to be too sensitive to be immediately useful. Another inherent disadvantage of this technique is that it cannot provide a positive result for girls who have X chromosomes like their mothers. Females can be recognized only by the exclusion of male DNA, which might lead to a lot of cases of mistaken identity. In fact, our colleague had been carrying a little girl all along. The march of progress will probably get around the problems of diagnosis, but at the moment such tests are dangerously misleading.

There is a technology already available for patients undergoing IVF treatment to choose the sex of embryos, although in the United Kingdom choice is permitted only where a genetic disease is carried in the male line. Preimplantation genetic diagnosis, or PGD, is used to select female embryos for transfer to the uterus. Sex selection for nonmedical reasons is forbidden, although it may occur inadvertently during laboratory conception. When embryos are left in culture for longer than normal practice (two days), the most vigorous ones reach the blastocyst stage by the fifth or sixth day, with the males often ahead of the females. By selecting either the most or the least advanced embryos, the odds of having a male child are increased or decreased by about 10%. However, we need not worry that IVF techniques, including PGD, will skew the sex ratio of the population, even if a strong preference existed for one sex or the other, because the overall number of babies conceived this way will always be relatively small.

X Certificate

A more acceptable method of sex selection would be to separate the sperm from a sample of semen into fractions containing mainly Y- or X-bearing sperm. The wanted type could then be used for artificial insemination. Splitting the semen has turned out to be much less reliable than expected, and it is still not possible to produce a string of one sex, like the parrot. Nevertheless, such has been the faith in these methods that about 65 sperm centers have sprung up around the world to offer a sperm selection service.

The London Gender Clinic opened in a suburban home five years ago. Inside, the patients discuss procedures with Alan Rose, a retired doctor. What used to serve as a kitchen–dining room has been modified to make a laboratory for Peter Liu, the biochemist who runs the spermatology. Their patients are mainly members of the 2 million British Asian community, and virtually all have come for a son. Whites, on the other hand, more often than not come for a daughter. The two staff members were taken aback when the medical profession joined the press in publicly condemning them, for they believe they are harmlessly serving the needs of patients and potentially bringing happiness into homes. There is, of course, a danger that parents who are disappointed that the method did not work for them might treat a child of the wrong sex differently or choose to have an abortion. Rose and Liu recognize this and try to obtain reassurances, even if the gesture is rather empty. But these worries also suggest the importance of developing a fully effective method for separating sperm.

The clinic, like others, pays an initial licensing fee and a royalty every time the technique is used to Gametrics, an American company founded by Ron Ericsson, who patented the sperm separation method 25 years ago. In a paper published in the journal *Nature*, Ericsson described how to produce an enriched specimen of male sperm, which he suggested swim faster than females because they are slightly lighter. The washed sperm are layered on top of a column of protein solution (albumin), and within an hour the male sperm are said to have sufficiently migrated ahead of the female fraction for them to be separated. Complete separation into the two kinds was never claimed, and in a recent study the average enrichment stood at 72%, but the results for female sperm were poorer. The specimens are used by insemination via a catheter into the woman's cervix or uterus at the fertile time of the cycle. The advantage of the procedure is that it is unlikely to harm the sperm and is relatively simple and inexpensive compared with IVF-based technologies. But since artificial insemination normally results in a pregnancy once in every ten attempts, the people who try this treatment know what it means to be patient.

There has been continuing claim and counterclaim about the efficacy and ethics of sperm selection. Most independent experts are sceptical, partly on technical grounds and because there are vested interests at stake in obtaining good results. When the method was tested in academic medical centers, there was no convincing confirmation. We tried to separate sperm and it worked—but only once and then never again. More recently, computer tracking has been used to test whether male sperm swim faster, but the only interesting finding was that females swam straighter. If the method did work, the ICSI technique for infertility might be expected to affect the sex ratio because sperm are slowed down in a thick medium before being picked for injection into eggs. But the latest results show that the sex ratio is little different from normal for babies conceived this way. It would be surprising if the method worked, given the enormous variation in shapes and swimming activity of sperm in a fresh semen sample.

But the incentives to find ways of separating sperm are great. Quite apart from private patients, there are huge profits to be reaped through helping farmers overcome the 1:1 rule in biology. The dairy farmer wants only cows and the poultry farmer only hens, with just the odd bull and cockerel for the purpose of insemination. It is surprising that what appears to be a simple challenge has turned out to be more difficult than making either transgenic animals or clones, though the efforts go on.

It was thought that there may be different proteins on the surfaces of the sperm that could be targeted. Attention focused for a while on a candidate called H-Y, which was presumed to be expressed exclusively on male sperm. Imagine the public reaction if women wanting only daughters could be immunized against H-Y sperm, and those wanting boys had their husbands' male sperm harvested from semen by binding to H-Y antibodies in test tubes. But this theory is probably flawed, because when X and Y sperm form from their progenitor cells, they do not completely separate until a late stage. Even large molecules can pass from one cell to another through large gaps in their membranes to equalize the amounts of RNA and protein, including H-Y. The last embers of communism may be dying out around the world, but they remain alive and well below the belt where these cells freely share their resources. This is a sensible biological arrangement—otherwise gender wars might break out between male and female sperm as they compete to be the best fertilizer.

Having ruled out differences between sperm protein and RNA, apart from the slim possibility of differences in sugars between male and female sperm, there is only one other potential difference to test for—DNA. Since the Y chromosome is much smaller than the X, the male sperm has less DNA per cell, though this amounts to a difference of under 3% in humans. Yet this difference is enough to separate the sperm, if not as efficiently as in farm animals, where the differences are greater.

Cells with different amounts of DNA can be separated using a sophisticated piece of equipment called a flow cytometer, which is a long tube containing a laser. After the cells' DNA is stained with a dye that is excited by ultraviolet light, they cross the laser's beam one by one. The glow emitted is amplified by a photomultiplier, converted from an analogue to a digital signal, and then displayed as a frequency distribution. The more DNA, the more light emitted. Once the amount of DNA has been estimated, timing and charging circuits are activated, and droplets carrying the Y sperm are given a negative electrical charge, those carrying the X sperm a positive one (a reversal of discrimination to please the gender police). They are then passed through a strong electromagnetic field where the droplets are attracted to plates with the opposite charge and, thence, to collection tubes—blue for boys and pink for girls.

The method works—up to a point. After sorting rabbit sperm using flow cytometry and fertilizing eggs in vitro, embryos were made and transferred to surrogate does. When the babies were born, 94% of those from X-sorted sperm were female and 86% from the Y sort were male. What is more, the babies were all normal, which is reassuring since the dye and the ultraviolet light are potentially mutagenic for sperm DNA. Progress is being made with farm animals, and even as I write, the results of the first successful clinical trial are being announced in the United States.

It has been difficult to adapt the methods for human sperm because of the smaller differences in DNA content. The technical problems are compounded because human semen is very heterogeneous and sperm have bullet-shaped heads that are difficult to resolve optically. The method falls short of full reliability. At the Genetics and IVF Institute in Fairfax, Virginia, where the method has been patented for clinical use, the separation is 60% to 70% accurate for male sperm and about 85% for female. Among the first 14 pregnancies for couples wishing to balance their families or avoid a boy with a sex-linked disease, 13 were female— as desired—and all were healthy.

The flow cytometer sorts sperm at a rate of about 300,000 per hour and soon generates enough for IVF or ICSI. Any IVF unit with access to the expensive equipment in its hospital could theoretically offer a choice of gender at less than $1,000 on top of the costs of IVF. The odds of having a daughter could then be raised from around 1 to 1 to 4 to 1, and the chances of having a boy would be just a little smaller. It has to be said, however, that this remains an experimental method, and, the quality of animal embryos has been found to be poorer after sperm sorting. I certainly would not choose it were I contemplating having a son or daughter until there was full safety awareness.

The big question is whether the technology will advance to the stage at which it can be used routinely and without the costs and inconvenience of IVF. Then we could imagine a service for any couple prepared to conceive by artificial insemination (AI) with the sorted sperm. The limitation at the moment is the large number of sperm needed for AI and the slow rate of sperm sorting, though speed is going up and capital equipment costs are coming down. Since most of the pregnancies in the Fairfax trial were established using AI rather than IVF, we may be closer to widespread application of this technology than had been supposed. As usual, a new technology begins slowly and tentatively but becomes more efficient with time. Attitudes will probably change, too. I think we will see greater acceptance of family balancing, and I doubt that it will ever be used or abused as much as is feared.

7

Other Wombs

Mother's Nest

The first nine months of our existence are the most formative in our lives. This is the time when mysterious processes knead the raw products of conception into a recognizable human shape and nurture the baby to full size within the womb. Pregnancy and labor are times of uncertainty, of risk, and of anxiety. That we start life inside another's body is one of the most extraordinary facts of our biology, although it is so familiar to us that we seldom stop to question the body's wisdom. How many brave mothers have died in pain during childbearing or shortly afterward from puerperal fever—and still do in many parts of the world? How many potential gurus and geniuses have perished in the womb or during delivery? So much human history has turned on the outcome of particular pregnancies and the inability of key figures to bear one. How different it might have been if we reproduced like turtles or in a glass bottle!

For generations, women had to accept that they would be the breeders and the men the hunters. Pregnancy was both a burden and a risk, but it was also their raison d'être. Women can now choose when and how many but not the physiological circumstances. The risks have diminished and sophisticated medical care and antenatal advice maximize the chances of a happy outcome. Pregnancy should be more enjoyable than ever before, but everyone knows it is also a huge upheaval; although the old concept of "confinement" is relegated to history, pregnancy still ranks as a major event in a woman's life. Women are starting to ask whether the cost is too high. Those who believe it is but still want to have children of their own may be warming to the idea of a substitute womb. Just as the natural breast gave way to wet-nursing and to powdered milk formula, could the womb one day be replaced? There are already some women, albeit still few, who would choose a surrogate mother for their social or professional convenience if they could. Is the next stage an artificial womb? This scenario may sound implausible or objectionable, but there are biomedical grounds, if not social ones, for welcoming it. Besides, what seems at first unnatural turns out to be less so when we consider our evolutionary origin and alternative arrangements for nurturing young in nature.

The platypus and the echidna provide examples of reproductive methods that other mammals left behind in the evolutionary past but that have proved their worth over millions of years. Instead of gestating their offspring in a uterus, they lay eggs, but unlike most egg-laying amphibians and reptiles, they do not simply lay and leave their eggs. They incubate them for ten days in a pouch, which is an external "womb" that develops in the mother's belly during the breeding season. It provides less protection for the brood than a uterus, but it is better than nothing.

Another Australian example, the red kangaroo, goes further in protecting its babies. After a pregnancy lasting less than a month, the joey is born weighing under a gram—compared to the 30 or so kilograms of the mother. At this stage its development is still rudimentary, but using its more rapidly developing front legs, it still manages to claw along a chemical trail in the mother's coat until it reaches the shelter of her pouch, where it clings onto a nipple for dear life. The joey remains here for the next six months, until the next tenant is ready to take up residence. All at one time, a mother may have an embryo in the uterus, a joey in the pouch, and another in the bush bouncing beside her—a very efficient system. There are plenty of examples of alternative modes of "pregnancy," and there is no standard length or form in animals.

Most mammals the size of a kangaroo have longer pregnancies and deliver larger babies that are less dependent on their mothers. Bigger babies are more likely to survive, though the mother pays the price in a longer gestation.

Sheep gestate for 5 months and deliver lambs that can frolic and graze right away. Elephants gestate for 22 months, and their calves are sufficiently well-developed at birth to keep up with the herd. Our species spends 9 months in the womb, and this is only the first part of a baby's long journey to full independence. The timing of our rude delivery into the outside world is crucial for the baby's survival, but it could have turned out otherwise. If the baby's lungs and circulation matured sooner an earlier delivery would be possible, and the responsibility for sustenance would switch from the womb to the breast at an earlier stage, as in the case of kangaroos. By contrast, a larger female pelvis, allowing a larger baby's head to get through the space, would make a longer human pregnancy possible. No one seems to be in favor of extending pregnancy, but since medical science has made considerable progress in saving premature babies through intensive care, the search has now started for a substitute for the womb.

The embryo arrives in the uterus on the fourth day after fertilization as a minute ball of about eight cells and floats freely for a few days, as if pondering the best place to implant. After two to three more days it is ready, burrowing into the uterine lining with aggressive enthusiasm and making a placenta immediately. The next few weeks are the most critical as it grows rapidly and cells move to their proper locations to form the future organs—the brain, heart, lungs, and so on. If anything goes wrong in the first ten weeks, or if there is a serious genetic fault, there is a risk that the pregnancy will miscarry or the baby will be born malformed. Just as a loaf cannot be reformed once a lump of dough is kneaded and baked, so an embryo, once it has started, cannot be repaired, though pediatric surgery makes valiant attempts to overcome disability and disfigurement after birth. Nothing is more awesome than the emergence in these early weeks of a recognizable human form from a tiny, undifferentiated mass. Such unique events might be expected to need a special environment, but the uterus is just a clever incubator, and rather less sophisticated than either the ovary or the placenta. Nature's real genius is being wrought within.

Just as a tropical fish survives in a heated tank on food and an occasional change of water, a fetus floating in its amniotic fluid is substantially independent of outside conditions. Its environment is warm, wet, and contaminated with its own waste products. The oxygen and nutrients needed for growth are obtained via an umbilical cord attached to the placenta, rather like the lifeline of a North Sea diver trailing from a ship. The placenta is a multipurpose organ and one of the miracles of nature. It performs the role of the lungs, intestines, liver, kidneys, and glands all in one until the baby can rely on its own. The placenta and chorionic membranes fix the territorial boundary between mother and child: no nerve cells cross and few blood cells migrate, and then only in the outward direction. This arrangement is a compromise between cutting the fetus off from the outside world and protecting it from being swamped by maternal chemicals. In principle, conditions for healthy growth and development could be provided elsewhere—even outside the mother's body.

Extrauterine pregnancy becomes more plausible when one considers that our evolutionary ancestors have opted for this strategy and that uterine pregnancy carries significant risks. If the mother dies or is harmed, her fetus suffers likewise, unless it is old enough to be safely delivered by cesarean section. Dire consequences can also result from early rupture of the fetal membranes and premature birth. Despite the wizardry of modern neonatal intensive care units, no babies born earlier than 23 weeks survive. Few of those that survive at this stage do so for long, and the risks of permanent damage to the brain, eyes, and other organs are high. Pediatricians will undoubtedly continue to push down the lower limit of viability, but extreme prematurity will always be dangerous until an effective substitute for the placenta can be found.

Uterine pregnancy also carries dangers that the fetus will become infected. Fetal tissue and fluids are almost ideal culture broths for microbes and must be kept sterile at all costs. Since the baby cannot be hermetically sealed, its safety from bacteria and viruses that get into the womb and cross the placenta can never be absolute.

The price of avoiding the microbial world in pregnancy is to be born immunologically naive and less capable of reacting briskly to fresh challenges from infectious agents. White blood cells and antibodies are the immigration officers and policemen of the body that keep the threats at bay, and several weeks are required to build up resistance to breaches of immunological security. Some of the mother's antibodies cross the placenta before birth and more are absorbed into the baby's bloodstream during breast-feeding until the baby can look after its own immunity. Nature knows best, to use a simplistic catchphrase, but this wisdom is fallible because harmful antibodies sometimes mingle with the beneficial ones. Antiphospholipid antibodies coming from the mother can cross the placenta too and are responsible for 10% of all miscarriages. In the past, many babies died or were severely anemic when a woman with a rhesus-negative blood group carried a rhesus-positive baby as a result of having inherited the father's gene. The first baby was always unaffected because the immune response was weak, but each additional rhesus-incompatible pregnancy boosted the mother's antibody response, aggravating the problem until prophylactic treatment came along with anti-D to prevent the problem from developing.

Nor is the placenta a perfect guardian of fetal nutrition, although it does a very efficient job of supplying the molecular building blocks of all tissues. It has built-in mechanisms for transferring salts, glucose, amino acids, fatty acids, and vitamins to the fetus, and water crosses by osmotic diffusion. The fetus can be harmed by a high sugar level in a mother who has diabetes that is poorly controlled and it releases more insulin from its pancreas in an attempt to bring its own levels back into balance. This reaction is as futile as King Canute's efforts to repel the sea, because the fetus's insulin can neither correct the hormonal problem in its mother nor stop the flood of her sugar. In fact, its efforts make matters worse for itself because insulin also encourages more fat to be deposited; a cherubic appearance in this case is not a healthy sign.

The placenta is a leaky sieve to most drugs. The evidence for pernicious effects is seldom as compelling as the thalidomide story, which caused limb deformities in 2,000 German and 500 British children born in the 1960s, but this tragedy has left a long shadow. Greater caution is now urged with prescription drugs during pregnancy than in almost any other condition. Likewise, concerns have grown about recreational drugs, which can also cross the placenta to affect the baby. The impact of binge drinking on the developing brain is so clear that the expression "fetal alcohol syndrome" has been coined. The baby can also fall victim to passive smoking. Carbon monoxide combines with the blood pigment to form carboxyhemoglobin, which compromises placental function and may suppress the mother's appetite. Babies of mothers who smoke during pregnancy are on average nearly 300 grams or 10% lighter at birth, which is a significant disadvantage. Yet, despite advertising campaigns and other warnings, patients dressed in gowns that barely conceal their swollen tummies are often to be seen standing and puffing smoke just outside our hospital.

A normal size at birth is a good omen for infancy and for later life, too. Examining old medical records from hospitals in England, David Barker, a public health specialist at Southampton, found that people weighing under 3 kilograms (6.6 pounds) at birth had ten times the risk of serious disease compared with those who were 4.5 kilograms (10 pounds) or more. The diseases are degenerative conditions arriving in middle to late life and include some of the greatest killers and causes of disability—noninsulin-dependent diabetes, high blood pressure, stroke, and heart disease—though apparently not cancer. We are still quite ignorant as to why many babies turn out small and why this should be a disadvantage in the latter part of their lives, though hormonal factors are probably involved.

Small babies tend to have lighter or abnormal placentas and are more often born to mothers with smaller frames. They also occur when a uterus designed for single occupancy is shared, which is the case for about one in four patients who conceive after IVF treatment. Twin and multiple pregnancies might seem "cute" and make economic sense after expensive fertility treatment, but they may cause heartaches later on. The babies are more likely to be born too early and have heart and lung abnormalities; triplets have a risk of cerebral palsy more than 40 times greater than average. It makes biological sense to have one baby at a time, though this is another limitation that artificial wombs would overcome.

Most of us could not have done better than our mother's womb, but sometimes there is a case for using another womb, and one day an even safer place may be found. Some women who have chronic medical conditions—heart disease, high blood pressure, kidney and autoimmune diseases, and severe diabetes—are advised against pregnancy. Other women seem perfectly healthy but despair of ever becoming mothers after repeated miscarriages or fruitless attempts with IVF. Yet more are unable to bear a child because they are born with a malformed uterus or without one or have had a hysterectomy to stop heavy periods or to remove fibroids or cancer. Hysterectomy is very common in the 35 to 44 age bracket, just at the time when increasing numbers of women are choosing to start their families. In the United States over the last 30 years, the overall frequency of this operation has risen from 25% to 60%, although the percentage is slipping back now. Those who have normal ovaries but cannot carry a child have to turn to other women for help for a full nine months.

Surrogate Wombs

Surrogacy is a biological contract between a couple who cannot bear their own child and a woman who offers to carry their child for them. It is not a modern invention, and it originally involved discreet insemination of the third party by the husband, whereas today artificial insemination or embryo transfer are more standard. In the Old Testament, Abraham's wife, Sarah, thought she was too old to have a child. Instead of just accepting that they would remain childless, she gave her Egyptian handmaid, Hagar, to her husband so that he could sire an heir. He gratefully accepted the chance, and some time later Ishmael was born. But when Sarah produced Isaac in her old age, her jealousy was aroused against Ishmael, the rival of her natural son for leadership of the family, and she drove Hagar and Ishmael out into the desert. Nowadays, primogeniture is rarely a cause of serious family quarrels, although tussles occur over child custody after surrogacy. In a highly publicized case, a British woman refused to give up her child to a commissioning couple from Holland. In the United States, the judge presiding over the Baby M case gave custody to the genetic mother, and the woman who bore the child was granted visiting rights only because she had "bonded."

Opinion is divided, but many doctors and counselors hope that keeping reproduction within the family by asking a sister or another close relative to carry the baby will minimize these sorts of risks. Surrogacy is a big thing to ask of anyone, but as kin have a genetic stake in the outcome, they are more likely to cooperate in an intimate biological contract. There have been several cases of older women—some of whom had even passed menopause—who have carried a child for their daughter who was able to produce eggs but could not go through the pregnancy. An unnatural arrangement is sometimes the only way to fulfil the natural ambition of having a grandchild, and would-be grandmothers are occasionally willing and fit enough to play the biological role in addition to helping with child rearing, which they have often done. Surrogacy, whether inside or outside the family, breeds strong emotions, and the psychological consequences can be unpredictable. Back-bedroom contracts, like the one between Abraham and Hagar, carry risks if they break a chaste marriage contract and introduce a third-party set of genes. IVF surrogacy, although often receiving unwelcome press attention, goes some way to avoiding these problems by recourse to technology. A woman who can produce eggs but cannot carry the child can undergo IVF with her husband's sperm and have the embryos transferred to a surrogate mother. The relationship is more secure because the baby carries the genes of the commissioning couple, and, because the gestating mother has no genetic interest in the outcome, there should be slimmer chances of conflict.

Embryo transfer was pioneered in animals a century ago. Whether embryos were transferred from pigmented to albino rabbits or in the reverse direction, the litters always bred true to the donors rather than to the hosts. It is a principle of biology that genes are inherited only from the union of the egg and the sperm, never during later stages of pregnancy. Embryo transfer is now routine in agriculture, and farmers transfer cow embryos from beef to dairy cows to produce an economically attractive ratio of stock for the market. In the United Arab Emirates, prize female racing camels breed without ever leaving trackside. Camel embryos created by IVF are transferred to flat-footed Egyptian breeds, which carry out the unglamorous surrogacy duties. Among humans, too, the surrogates may be poorer, more disadvantaged individuals than the couples who commission the pregnancy, which is one reason why commercial surrogacy has not been permitted in the United Kingdom.

One of the curious facts of reproduction is that the uterus does not concern itself with the embryo's origin and, provided the organ is healthy, one serves as well as another. A woman's body will reject an organ transplant from an unrelated person, but her immune system does not affect the placenta, even though it carries genes derived from the father. Were the uterus less tolerant, pregnancy would never have evolved and we would still have to lay eggs. Consequently, there is no need to find a compatible match between the embryos and the uterus. For successful implantation the main precondition is synchrony of the embryo and the stage of the menstrual cycle in the surrogate. There is only a brief window of opportunity for implantation in the uterus, which occurs 20 to 24 days after the beginning of the last period; if it is missed, the embryo perishes.

The first surrogacy attempt using IVF and embryo transfer was made by Patrick Steptoe and Robert Edwards at Bourn Hall Clinic, and it was greeted by a furor in the media. They were approached by a woman who had normal ovaries but no uterus, and they could find no reason to refuse help. This case was unsuccessful (only 40% are successful), and the first baby produced by IVF surrogacy did not arrive until 1989, and then in Iceland. Now, almost any woman who can produce an egg can become a genetic mother, providing she has both a male partner and a female surrogate to cooperate. Like an actress who is temporarily relieved by an understudy, the woman who supplies the egg wants to get back on the stage as soon as possible and play the full role of mother. Like any audience though, the woman's family is likely to mutter about the plot and the choice of understudy.

The rest of society takes a keen interest in these arrangements too. Attitudes of governments and the world's major religions range from liberal to reactionary. In the United States, family matters are regarded as "states' rights," and some odd policy combinations have emerged. In Louisiana, for instance, surrogacy is permissible but egg donation is not. Advertisements for "wombs for hire" in newspapers and over the Internet have increased public concern and could frighten off women who are motivated more by compassion than by profit. The fact that surrogacy is used by the homosexual community adds to the controversy. The Surrogacy Arrangements Act (1985) in the United Kingdom permits surrogacy services but forbids them for profit, apart from "justifiable expenses" for pregnancy. It seems incongruous that professional staff are permitted to earn fees for their small part in the contract, but the surrogate woman is not. In Britain, IVF surrogacy costs about £10,000, compared with $50,000 or more in the United States, but a lot of natural surrogacy goes unrecorded and the charges are never known.

IVF surrogacy is still uncommon, and only about 50 babies have been born to date after this treatment at the largest service in the United Kingdom, Bourn Hall Clinic, near Cambridge. There is no "standard" couple seeking surrogacy, and, with few exceptions, the only common denominator is the inability to bear a child for health reasons or the lack of a uterus. Models who wish to dodge pregnancy for aesthetic reasons and professional women who want to avoid interrupting a career are turned away by the clinic. Surrogacy sometimes has bizarre twists, to the advantage of headline writers and much to the irritation of cautious practitioners. In Italy, a woman became pregnant with twins, each from a separate set of parents. The combination could have involved even more parties if donor eggs or sperm had been used, but such webs of relationships in blood and law are unnecessarily complex and most cases are a good deal more straightforward. Most couples who want another child repeat the arrangement with their first surrogate, but one British family was forced to engage separate women. The first delivered a healthy baby but in the process hemorrhaged so seriously that she had to undergo a hysterectomy, like the woman she had acted for. This unlucky case illustrates why volunteers should complete their own family first.

Not every healthy woman is suitable for surrogacy, and the selection process can be a harrowing experience. There are arguments for and against using family members or friends, for a triangular relationship can enrich as well as destroy. On balance, a contract within the family is more attractive than one with an unknown person, and in almost a third of cases a sister, sister-in-law, or mother/stepmother volunteers to help. But however close the relationship and committed the parties, the contract is emotionally charged, and plenty of careful counseling is needed on both sides, with time to consider all possible outcomes. Who takes the blame if anything goes wrong or the pregnancy ends in miscarriage? Will the carrying mother feel guilty if it does? What happens if the baby is born defective? Has the surrogate taken precautions to avoid getting pregnant otherwise? If she smokes or drinks will she give up during pregnancy? These are some of the worries that can surface during the counseling and contracting stages. The doctor's responsibility is to give the best chance of success at the least risk. After IVF treatment the embryos are stored for six months until the surrogacy partners have been screened for hepatitis and HIV, and only when the all-clear has been given will the embryos be thawed and the pregnancy begin.

Some parents wonder whether their choice of surrogate will affect their baby's characteristics. If natural surrogacy involving sexual intercourse is involved, the answer is obviously yes. The woman who bears the child will have contributed the egg and, therefore, half the baby's genes. With IVF surrogacy there should never be any doubt, for the genes of the child were decided at fertilization and the host's cells never cross the placenta. Whether one woman's womb is better than another's and might produce a brighter and more beautiful baby is a harder question to answer because we cannot experiment with human beings in this way. Taking an extreme example of transferring the embryo of a Shire horse to a Shetland pony, uterine capacity obviously affects the size of the newborn foal, regardless of its genetic origin, although it catches up with normal growth after birth. There's no evidence that the physical appearance and IQ of a child are affected by the choice of a healthy surrogate who avoids taking anything harmful in pregnancy.

It is theoretically possible—although no more than that at present—to provide a replacement uterus for women who have had their own organ removed or have been born without one. I have received requests from far afield for news of uterine transplant research, even though until very recently it was hardly dreamed of. If a healthy organ is donated by a woman undergoing a hysterectomy, and— here's the difficulty—it is not rejected by the immune system of the woman who receives it, there may be a slim chance of successful pregnancy. The feat would require skillful surgery to join the blood vessels to the transplant, and probably IVF would be needed to start the pregnancy and a cesarean section to safely deliver the infant. On balance, the risks seem unreasonable when surrogacy is a safer and more certain means of producing a child. But if we could be sure of its safety, pregnancy in a transplanted uterus would have the great advantage of keeping pregnancy within a marriage, avoiding all the social problems that surrogacy can pose.

Devoted Fathers

Researchers for the Hollywood spoof *Junior* approached some of my clinical colleagues for advice on how to make the film's leading actor, Arnold Schwarzenegger, "with child." This prompted some lively discussion in our obstetric unit. The movie has been provocative at a time when more fathers are sharing domestic roles and bringing up children as single parents. If fathers could also share the job of childbearing, it would be the ultimate liberation for women from their biological chains. But most debates have focused on the parents and their relationship rather than on the child, and Junior hardly featured in the film at all. A child's interests are unlikely to be served by male pregnancy, at least in the way that it is currently conceived, if you will pardon the pun.

What we regard as conventional sexual distinctions are sometimes overruled in nature by alternative social arrangements if they enhance reproductive fitness. In some species of fish and frogs the males incubate their young in their nest or even in their gullets, which they can do as well as any female. Male birds in some species draw the short straw. Male phalaropes sit on the eggs laid by the females who fly south from Iceland to warmer climes. And in Australia, the male Mallee fowl is responsible for tending a nest like a compost heap to control the temperature of incubation. God may have "created them male and female in the beginning," but a lot of differences in domestic relationships have emerged since. Perhaps He would not disapprove if men shared the biological burden of pregnancy, and I guess some female partners would not refuse the offer. But setting aside curiosities from the animal world, is male pregnancy in humans possible or just a twinkle in a movie director's eye?

The fictional story of Junior was based on the fact of ectopic pregnancy in women, which is far more common in our species than any other. At a frequency of about 1 in every 300 pregnancies, it is something every doctor encounters, though it occurs ten times more frequently in countries where pelvic infection is commoner. Ectopic pregnancy usually occurs in the fallopian tube where a blockage halts the journey of the embryo to the uterus. Much less commonly (1 in 10,000 pregnancies) the fetus becomes implanted in the abdomen because the tube or uterus ruptures or the embryo has floated off into inner space and rested on an abdominal organ. Some ectopic pregnancies come to nothing and most are detected early and terminated surgically. Occasionally, they go to term and a baby is delivered by cesarean section, which, in the case of abdominal pregnancy, should be managed by the most experienced member of the obstetric team, because if the mother's blood vessels to the placenta are torn she may quickly bleed to death. After the water has broken, the baby is lifted out very gently and its umbilical cord is tied and cut. In contrast to what happens normally, the placenta of an abdominal pregnancy is left behind to wither in situ.

So desperately does the embryo cling to survival that it can implant almost anywhere and at any time. The trophoblast cells around the embryo that will form the placenta are exceptionally invasive in humans and, like cancer cells, can bury themselves in almost any tissue. The uterus staves off the invasion to balance the fetus's needs against the mother's interests. What a pity it is not as effective in protecting against the spread of uterine cancer cells as it is in policing the trophoblast!

Given the facility with which embryos can implant, doctors could probably establish pregnancy in a male belly after injecting an embryo through the skin with a syringe, and the consequences could be predicted from what happens spontaneously in females. It is unlikely that every embryo would implant, but those attaching themselves to a favorable site at which they could attract a good blood supply might survive. Within the first month, long before the belly started to swell, the father's urine could be tested for the pregnancy hormone HCG. And as men are as sensitive as women to the hormones estrogen and progesterone, which are produced in large quantities during pregnancy, changes would occur in parallel. The testicles would shrink, just as a woman's ovaries are suppressed in pregnancy. Hormone action would encourage fat to accumulate on the male thighs, buttocks, and breast. In the movie the star kept his macho chest for the sake of public taste and his reputation, but the filmmakers did make one concession for the sake of plausibility—Arnie cried. Along with a bigger appetite, the pregnant man would experience food fads and constipation, nausea and vomiting. Not all of the changes are hormonally driven, of course. Given some encouragement, he would probably succumb to the temptation to gaze into baby supply shop windows to complete the reversal of roles.

Male pregnancy might seem attractive to some women, but I doubt whether it would ever serve Junior's interests. The risk of birth defects in ectopic pregnancies is 50%. Outside the thick, muscular walls of the uterus, the soft tissues of the fetus are unprotected and easily compressed. Whether an abdominal pregnancy would be any worse in a man than a woman is hard to say, though there is greater concern for a female fetus than a male one. We used to regard the sexual development of the fetus as inviolate because maternal hormones are turned away or tempered before crossing the placenta. This tidy theory received a knock when it was found that a female rat fetus flanked by two males became slightly more androgenized than one between a sister and a brother, and much more so than one between two sisters. Although there is no evidence of such hormonally induced changes in human multiple pregnancies, the question remains whether in the case of a pregnant man his male hormones might be high enough to affect a female fetus.

Next time around, the makers of *Junior* ought to try uterine transplantation. This procedure could be more effective and safer than an abdominal pregnancy for both father and child. The big obstacle is the danger of the father's immune system rejecting the organ. But if this could be overcome without the use of potentially harmful drugs, male pregnancy would become a distinct possibility. The experiment could theoretically occur in nature where a fault during development had caused both male and female organs to form in the same individual (i.e., hermaphroditism).

According to a 1997 report, doctors in Meerut in northern India found ovaries and a uterus in a 35-year-old man during a routine operation on his stomach. He had been married for ten years, although his wife had not borne a child, and neither, needless to say, had he. The examining doctor believed that his uterus could have sustained a pregnancy if an embryo had been transferred after IVF. Intersex farm animals that possess organs of both sexes are well known in veterinary circles, though a vital part of one or both of the sets is usually missing, and I know of no instances of hermaphroditic conception. We shall never know whether the man (that was the gender assigned to him at birth) could have undergone a pregnancy because his female organs were removed and pickled for exhibition.

A true male pregnancy is one of the few theoretically possible variants of the reproductive process that has not yet been recorded. Considering the availability of safe alternatives, there is no case for engineering a male ectopic pregnancy. Likewise, there is no justification at present for even dreaming about growing babies in bottles, even in special circumstances where a less hazardous alternative to gestation in the body is needed.

Babies in Bottles

"One by one the eggs were transferred from their test tubes to the larger containers; deftly the peritoneal lining was slit, the morula dropped into place, the saline solution poured in . . . and already the bottle had passed and it was the turn of the labelers—heredity, date of fertilization, membership of the Bokanovsky Group . . . the procession (of bottles) marched slowly on . . . into the Social Predestination Room." Aldous Huxley's imaginary world of baby farming is so graphic in technical detail and redolent with political innuendo that we can no longer regard the making of babies outside the body—or ectogenesis—with complete objectivity. In *Brave New World* the motive for using artificial wombs was to tailor the characteristics of babies for social ends rather than to liberate women from pregnancy or to avoid disability or disease. That is why the scenario fills us with so much horror. But were the technology developed with the health and safety of mother and child exclusively in mind, it could become a welcome advance in the twenty-first century.

Huxley was a well-informed layman. His elder brother, Julian, was a famous biologist, and the geneticist J. B. S. Haldane was a family friend. In Huxley's day, the science of human reproduction was new and the stormy dawn of ART was still far off. Knowledge of human embryology was rudimentary, gestation in the uterus making research so much more difficult than in the case of egg laying species. Huxley cleverly wove into his story of production-line biology the political and social anxieties of the 1930s. In extending the "industrialization" of reproduction in humans, he had plenty of examples of "factory farming" of reproduction in both commerce and nature to draw on. Battery hens produce hundreds of eggs of predetermined size and quality annually. Queen bees lay millions of eggs whose fate is set by the diet the grubs are fed when they reach a critical stage of development. In combining ectogenesis with cloning Huxley hit on the ultimate technology for improving and controlling nature. Of course, he realized that wombs were unlikely to be replaced by glass vessels for a long time. But he would have been amazed to see the progress ART has already made and to find that ectogenesis is now being discussed as a plausible technology for the future.

Artificial wombs would put an end to all the medical and social difficulties of surrogate motherhood, so that conventional maternity would become unnecessary. With a bottle for a womb and culture fluid as a substitute for maternal blood and amniotic fluid, it would be possible to complete a process in vitro that began at fertilization. The risks of maternal death, which in Britain were around 1 in 200 births in Victorian times and are now 1 in 10,000, would be eliminated at a stroke. Whether the technology would be too much of a luxury to be afforded in the developing world is an issue, for it is there that maternal health problems are most acute and the benefits would be greatest. Throughout much of Africa and South America a woman has a lifetime risk of dying in pregnancy or childbirth of between 1 in 25 and 1 in 50, reflecting both poverty and the low priority of women's health.

There would be advantages for the baby as well. Safe in its bottle, it would be spared any harm arising through the ignorance or carelessness of the parents—although accidents can happen even in the best laboratories and the level of care required by laboratory staff would have to be even higher than in ART. Currently, embryos are usually returned to the body about two days after fertilization, or slightly later, as a little ball of cells, long before they reach the critical period when the brain and other organs are formed. The early stages of embryo development are more resistant to developing abnormalities because formation of the organs is yet to come. Later stages could more easily go awry in vitro, and great care would have to be exercised to ensure that the best conditions were provided for the full 38 weeks of gestation.

The sterilization of culture vessels, instruments, and culture fluid should avoid any risks of passing on pathogenic bacteria or viruses from mother to baby in a normal pregnancy. Nor would any harmful maternal antibodies be passed. Furthermore, the nutritional needs of the baby could be delivered at the optimal rate, avoiding any maternal dietary fads or metabolic disorders. Perhaps most significantly, pregnancy in a bottle would make it possible to monitor the fetus at every stage of growth. At present, we can observe an embryo for only a few days after IVF and then, once an embryo has been transferred to the uterus, everything literally goes dark. If a fetus was growing in vitro, any abnormal development would be detected immediately and a quick decision could be made whether to treat or terminate it. Surgery in utero is something that rarely is contemplated at the moment, even for the most precious fetus, but it would become practicable at last if the fetus were in vitro. Any incisions made in the skin of a fetus heal quickly and leave few if any scars. Taking blood samples for lab testing, injecting drugs, hormones, and antibodies, and even transplanting cells or organs to meet a medical need would all be easier.

It should hardly need stressing that the fetus should be respected and never used as an object for casual or risky experimentation. Any temptation to interfere in ways that are not in the interests of the baby should be firmly resisted. But ectogenesis would provide a great opportunity to increase knowledge of what is one of nature's last great secrets, and it would greatly benefit fetal medicine in general. There is perhaps no subject in biology that fills us with greater awe or of which we are more ignorant than the molding of a baby in the womb. Some people may prefer us to leave nature alone, as they did when antisera and organ transplants first became available for treating ailing children and adults, but the chance at last to understand the most tender period of existence and, even more important, to cure diseases and help with the creation of life will surely prove irresistible.

I do not expect to witness ectogenesis in my lifetime, but I believe that when it is possible, the experience will enhance rather than diminish the value we place on a fetus. If watching a tiny embryo develop to the eight-cell stage is still a source of wonder, then the fetus's continuing story will provide an even greater marvel. But it is difficult to know how parents might react to extrauterine pregnancy. Mother and father would have to adjust to a more equivalent role. Without a baby growing and kicking inside her, the mother might feel less bonded, but the father might feel more so as he would be able to watch his baby developing. Any anxieties about the circumstances and timing of labor would be allayed as premature birth and labor pains became things of the past. The time of delivery would be decided by doctors and technicians when the ideal weight was reached and the heart and lungs were sufficiently mature for the baby to emerge from its bottle into the outside world and make its first cry.

The main justification currently for pursuing this research is not for the social or professional convenience of women but for the sake of high-risk pregnancies; it would offer a means of rescuing babies in the cesarean tradition. But at present it is extremely difficult to grow human embryos in vitro for 1 week, let alone 38. A few bold scientists have kept animal fetuses alive ex utero for a few days, although even this modest feat is not easy.

No one has succeeded better in this than Denis New, who worked in the Cambridge laboratories alongside Robert Edwards when IVF was being pioneered. As a young researcher in the early 1970s, I shared with his staff an office full of bottles of rat fetuses rotating like passengers on a miniature merry-go-round. The fetuses were recovered from the uterus shortly after implantation and bottled in culture fluid. They grew at a normal pace from the eighth day to the twelfth day after fertilization and met all their developmental milestones on time. They formed a brain, a beating heart, and paddle-like limbs. Everything seemed to be going fine and should have proceeded as far as "birth" on the twenty-second day. But then they stalled at the halfway stage.

The technique had worked up to day 12 because the fetuses obtained their supplies of oxygen and nutrients by diffusion from the culture broth through a special membrane called the yolk sac. But as the fetuses continued to grow their sacs failed to meet demands and a proper placenta (allantois) was needed to take over, but this did not happen in vitro. It was frustrating to see the fetuses form all the major organs but not achieve the seemingly easier job of growing to full size. The biological equivalent of a moon landing had been achieved, but going to the end of pregnancy still looks like a distant planet. If this is so difficult in a species with a placenta weighing less than a gram and that gestates for less than a month, how much harder would it be in our own species, which needs a placenta weighing half a kilogram that has to last for nine months?

If I were a gambling man I would bet on a marsupial being first past the post in the race to finish an extrauterine pregnancy. This is because their fetuses stick with yolk sacs throughout pregnancy. They develop rapidly and are born very immature, so they do not need the sophistication and efficiency of an allantoic placenta. Much progress has been made by the Australian biologist Lynne Selwood, who is studying the stripe-faced dunnart, a creature about the size of a house mouse with a pregnancy lasting just 11 days. Selwood has managed to culture its embryos to within a day of birth, so the race is almost won.

Considerable ingenuity will be needed before we can move far from the starting block in the case of humans, and most biologists are intimidated by the difficulties. The new method is unlikely to be the answer because the human yolk sac shrinks early in pregnancy and cannot provide all the functions of the true placenta. The placenta may look like a homogeneous lump of meat, but it is, in fact, an amazing one-man band serving its fetus's interests in all sorts of ways.

No one has yet produced an effective substitute for this organ, regardless of press claims about the much-vaunted "artificial placenta." A plumbing device invented in Japan for helping goat fetuses survive outside the womb may eventually lead to ways of helping premature human babies survive longer in intensive care units. But something much more sophisticated than tubes and pumps will be needed to replace the biochemical dialogue between mother and fetus for full development, and our inventions are likely to be a good deal less compact than the real thing. Hidden within the placenta is a folded membrane with a surface area measuring more than 100 square meters that allows the fetus to absorb the goodness from the mother's blood and despatches waste products in return. This is much larger than the fictional membrane that Huxley squeezed inside one of his bottles, and the challenge of making an effective artificial placenta is still stumping the most inventive bioengineers.

Advances in science proceed unevenly, and what seems difficult sometimes turns out to be easier than we supposed, and vice versa. We can detect a change in a single letter of a gene, but we cannot yet make a substitute placenta. We shall see more amazing progress in genetics and embryology in the next few years, but we will have to continue depending for some time on that "simple" muscular bag that we call the uterus for having our babies. This organ, though imperfect, has served us well. Few people seem to be clamoring for ectogenesis, which is a futuristic idea forbidden under most countries' laws. It is the most ambitious and least hopeful of the technologies I have discussed so far, and scientific progress has slowed to a crawl. Carrying forward the process started with fertilization and continuing human development in vitro through the stages of fetal development to the full-term infant is the greatest technical challenge in ART. Not for 100 million years—since embryonic membranes were first adapted to make a placenta and sweat glands modified to produce breast milk—would a mammal have undergone a bolder evolutionary step. As with other revolutionary changes in reproduction, the arrival of ectogenesis would probably herald a host of new opportunities for our species—social as well as biological.

8

Never Too Late?

Having It All

After World War II, men and women returned home to civilian life and set about making babies with a vengeance. Parental instincts, which had been bottled up during the Depression years of the 1930s and the war years of the 1940s, were given free rein. Not only was the world more peaceful, it was more prosperous too. There was full employment, rationing of scarce goods in Europe was tapering off, and Britain had declared itself a Welfare State. Not for many a year had there been more favorable omens for settled existence, and greater numbers of men and women than ever before were marrying and starting families.

This was the baby boom, and my brothers and I were some of its products. The boom lasted from 1946 to 1964 and, apart from the agonies of Vietnam and the tensions of the Cold War, fortune was kind to the generation that grew up and flowered in the 1960s. Most western countries were spared conscription, poverty and hunger, and most of the epidemic diseases. We enjoyed better public health, more employment protection, and freer access to higher education and the professions than ever before. This was a golden age of opportunity and abundance, which, in the past, only the most privileged in society had enjoyed.

Female baby boomers made even more gains. With the doors of opportunity springing open and the pill in their handbags, women were able to aspire to life styles and careers more ambitious and independent than their mothers had ever dreamed of. By keeping the reproductive anchor on board, some sailed into glittering careers, deferring fertility to some distant port and until the right man came along. These are the women who are considered successful today, whereas in the past it was those who quickly settled down with a lifelong partner to have children. But as the boomers approach and pass through their forties, they are turning to one more ambition to complete a well-rounded life—children—only to discover that their biology has not adapted to the changes in society.

These are the women who, with their partners, fill the waiting rooms of infertility clinics across America and to a lesser extent Europe. Society has encouraged them to excel in education, to launch a professional career, and to build a financially secure future, but it has neglected to help them invest in reproduction at the age nature intended. Other couples attending clinics may have been less imprisoned by the call of a career but did not find the right partner until middle life, or have remarried and decided to try for a second family. They may have a perfect home with a paid-off mortgage and the means to afford a nanny and private education, but they cannot get pregnant. These are the straits that many of my generation have found themselves in, and there are probably more middle-class people who are still childless than ever before in history. For some this has been their choice, while others still bide their time hopefully and still more look back regretfully.

Perhaps we are more aware than ever before that biology does not tarry and that female fertility moves inexorably downhill from the prime years in the early twenties to extinction at menopause. It seems unfair when society's reward system encourages the postponement of a family that the batteries of the ovarian clock run down just as a couple is ready for children. Women can now expect to live to nearly 80 years old compared with only 44 in the early Victorian era, yet the age of menopause has probably hardly changed since we were dining on mammoths in caves. Biological changes take place very slowly compared with social ones, and desperate measures may be required to bridge the gap.

Our attitudes to those who choose to remain childless until mid-life tend to be ambivalent. People who have not enjoyed professional success may feel that no one should expect to "have it all." And to many it seems foolish or just plain selfish for a woman to begin a family in or after her late thirties. The older a woman is the more firmly these convictions are held, and a woman in her fifties trying to cross back over the menopausal Rubicon with the helping hand of science can expect almost universal condemnation. Needless to say, when a man of the same age fathers a child with a younger woman he wears his achievement with pride and is heartily congratulated. He only has to bear, at most, jokes in poor taste about the virtues of Viagra and the hazards of cuckoldry.

Aristotle advised "setting the marriage age for girls at 18, for men at 37, or somewhat less" and set a trend in attitudes that has persisted ever since. The medical profession is as skeptical about older mothers as is society in general and still regards anyone over the age of 35 as gynecologically geriatric. Those over 30 bearing their first child are classed, rather pejoratively, as "elderly primigravidas." Most fertility clinics set an upper age limit for women but rarely for their partners. This has more to do with health economics and professional prejudice than with technology, for almost every woman who is healthy and still in possession of a womb can be assisted to become pregnant. In Britain, IVF treatment for older women is difficult because they are usually excluded from treatment on the National Health Service (in districts where it is available) and there is an overwhelming shortage of egg donors. Where eggs are available treatment can be as effective for them as it would be for women 10 or 20 years their junior. They don't even have the option of turning to adoption agencies, at least in the United Kingdom, because the upper age limit is set at 35 (currently under review). So by the time couples reach what they had thought would be the prime of life, they find they have been stigmatized with attitudes usually reserved for people more than twice their age.

More than 2,000 infertility specialists attended a pan-European congress in Greece in 1994. After a lively debate on "This house approves of using assisted reproduction for postmenopausal motherhood," the delegates decisively voted the motion down. Contrary to the tabloid image of doctors and scientists always pandering to the want of patients, in practice they reflect the same spectrum of conservative and liberal opinion as in most other walks of life. Even some of the pioneers of fertility research voice doubts about treating women past the natural age of menopause of around 50. Robert Winston, Britain's best-known doctor, who was present at the debate, has stated in one of his books that it is wrong "to subvert a natural biological event. To do so seems to me to debase the value of menopause."

A Gallup Poll carried out for the *Daily Telegraph* in 1997 revealed that the majority of British people also disapprove of fertility treatment for older women. A full 90% believed that women should stop trying for children between 40 and 50 years of age, although nearly half preferred to leave the decision to the patient herself in consultation with her doctor. Our ageist society has a short memory. Only a few generations ago many women were still breeding in their forties. In the 1920s, the average age of last confinement in Britain and America was over 40, and the record age for natural conception stands at 56 years. Surely it would be more logical to draw the line for assisted reproduction at this higher age, at least for the healthiest of would-be mothers.

A meeting of fertility experts convened in 1998 at the National Institutes of Health to discuss medical concerns and to set research agendas for the American "epidemic" of late motherhood. We were told that cesarean sections, pregnancy diabetes, excess weight, and hypertension were all many times more likely in mothers of mature years. Also more babies were born too small or too large or too premature. According to the medical pessimists, a mother over 35 is a high-risk patient, and an over-45 is a disaster waiting to happen. We were told that rather than expressing joy at the news of a positive pregnancy test, a doctor's reaction may be to darkly hint at termination even before prenatal screening has been done for the "sake of the health of the mother."

This list of despair should be taken with a pinch of salt, for medical minds are trained in averages but diseases and medical disasters occur as probabilities. Everyone is different, and our health and vigor decline with age at very different rates. Some people develop early the problems of old age, such as diabetes, heart disease, and bone loss, whereas luckier ones stay fit well into their seventies and eighties. It would be surprising if fertility were any different. Some people have the "change" at 45 and others at 55, some are fitter at 50 for pregnancy than others are at 30, although we put up obstacles only for the first.

People should always be treated as individuals whatever their age, and those who are fit enough ought to be given the green light for fertility treatment. When Mark Sauer, then at the University of Southern California, provided egg donation for superfit women between 40 and 52, he provoked fury, even though the pregnancies turned out to be successful and had few complications. Nowadays, treatment of this age group in American IVF units no longer draws so much comment. The age of contention has risen to 60, and the world record for delivery now stands at 63. Some people weather better than others, and there is a connection between reproductive fitness and all-around health. Those who have had a late baby naturally or a late menopause are more likely to live longer and join the salon of centenarians. Since lives have become healthier and longer, a woman giving birth even at 60 can expect as many years with her child as her grandmother had with her last one.

The quality of those years count more than the quantity, but few people have taken the trouble to discover the experiences of older mothers. Julia Berryman, a psychologist at Leicester University, has conducted a rare survey of hundreds of women contacted through appeals in newspapers and women's magazines. A group of women aged over 35 at their first delivery was compared with a group of women of the same social class who were in their twenties at the time of first delivery. The results were reassuring. Most of the older women were "very happy" and nine out of ten recommended the experience. Pregnancy was no more bothersome, though there were the usual irritations—heartburn, swollen feet and ankles, unruly bladders, and aching backs— and the older women were more likely to suffer extreme tiredness (although not all do). Despite few complications, more were signed up for cesarean sections by doctors who were skeptical that their uterine muscle had enough athletic prowess to manage labor. Fewer older mothers declared that they had bonded with the baby by mid-pregnancy, perhaps because they had miscarried in the past and they regarded pregnancy as a more uncertain condition or perhaps because they were worried about having a fetus with Down's syndrome, a worry that would have been reinforced by the greater attention on pregnancy screening.

Most parents put above all else the interests of their babies, who need more than a healthy womb and a full breast to get a good start in life. The Leicester survey found that late motherhood provided just as much love and devotion and in some respects was more advantageous. The older mothers had more positive attitudes to discipline, resorting to physical punishment less frequently in the training of their children, and they made more earnest attempts to breast-feed. The children were encouraged to be more independent, perhaps because the mothers and fathers were more self-confident after years of working. Yet the infants were evidently not shortchanged, for they acquired higher scores in verbal skills by 4 years old, were more self-motivated, less easily distracted, and more intellectually advanced (though not necessarily more intelligent). Older fathers spent less time at play but were more likely to sit with their child reading stories or watching TV and reliving memories of the receding days of their own childhood.

Perhaps we should not be surprised that baby boomers are in earnest about giving their kid(s) a good start in life and preparing them for the competitive world ahead. Concerns about a bigger generation gap than in earlier decades are probably unwarranted and would be puzzling to many cultures. Children in West Africa and other parts of the world are traditionally raised by their grandparents. A break with custom always triggers anxiety, but, put into perspective, older families have slight impact on society compared with the steady dissolution of the traditional extended family, which is now far advanced in the West. Besides, the children of older parents are more likely to have all their needs (and more of their wants) met and, because these families tend to be smaller, are unlikely to suffer from want of attention.

The edging up of maternal age is hardly perceptible from year to year, but with hindsight a minor revolution has occurred. Most American births are still to women in their twenties, but the average age for starting a family has crept up to 28, and 10% of births now occur after 35, with similar trends in Europe among the middle classes. The percentage of women aged 30 to 34 years having their first child rose from 7% to 22% between 1973 and 1994, and the increase was even higher for those aged 34 to 44. Many of the late maternities are to middle-class and better-educated couples. After the age of 40 more pregnancies now lead to delivery than termination, which says a lot about attitudes, and the women seem to be prouder of being an older parent and to care less about being mistaken for a grandmother collecting a child after school. When left to their own devices, people invest in reproduction according to what they perceive as their best interests.

Whether this drift to later motherhood goes much further will probably depend on perceptions about security and the economy. In Chapter 2, our species was typed as the most *K*-selected species because, living long in a secure environment, it can afford to breed infrequently and later. But how much later must be finely judged to avoid the infertility trap of mid-life. What couples need to know is how late they can wait to conceive and what can be done to improve the chances.

If women could hear their biological egg timers beeping, they would be able to plan their family with fewer fears of being caught out by an unexpectedly early menopause. A history of treatment with some types of chemotherapy or ovarian surgery increases the likelihood of an early "change," and women whose mother or sisters had a natural menopause before 40 should be aware that they may have inherited the tendency. One day it should be possible to screen young women genetically for premature menopause, though not to predict the month of their last menstrual cycle. To obtain that precision it would be necessary to count the total numbers of eggs in both ovaries, which is currently impossible using scanning methods. But if we knew that, say, 25,000 eggs remained, about 16 more years of menstrual life would be expected, if 4,000 then only 2 or 3 years. This information would help couples decide when to put their reproductive skates on, as well as providing confirmation for a woman with irregular cycles of whether she is already perimenopausal.

If a biopsy were taken from the ovary, like a blood test for anaemia, could "ovarian age" be estimated? Imagine a doctor using keyhole surgery to collect small pieces of tissue and sending them to the lab. Although the numbers of eggs could be counted under the microscope, estimating the total in both ovaries from the sample is unreliable. In a blood drop, there are huge numbers of cells, and every drop is very similar. But there are far fewer eggs in a biopsy and, since they are usually distributed unevenly, one sample may be substantially different from another. This is a pity because egg counting could forecast menopause decades ahead. Hormonal methods that are in current use can confirm that a woman is perimenopausal but there is still a great deal of uncertainty.

Hormone levels in the bloodstream start to change ten years before menopause, or around 40 years of age or earlier if the egg store is low. Hormones are the language used for communication between the ovaries and the pituitary gland, a pea-sized organ lying between the bony palate and the brain. The pituitary secretes several hormones, but the fertility hormones, FSH and LH, are the most important ones for the ovaries. They help the follicles ripen and produce estrogen and trigger ovulation when the time comes. Estrogen and another ovarian hormone, inhibin, tell the pituitary how things are progressing and when to turn down production of FSH to avoid too much stimulation. The same principle is used in the contraceptive pill, which uses both estrogen and progesterone to shut down the fertility hormones.

The dialogue between the ovaries and the pituitary, together with the upswings and downswings of their hormones, goes on month after month. Often there are no major changes in rhythm until the late forties. But the waste of follicles that has been all this time continuing in the background eventually causes the hormonal "voice" of the ovaries to fall. The pituitary gland senses that the conversation is becoming strained and raises its volume, as if speaking to a deaf person. The effort is in vain because, despite increasing FSH and LH production, the ovary is unable to reply as it used to. The frustrated pituitary even tries different hormonal "dialects" to get through to the deaf ovaries and begins to make its hormones more acidic and longer acting so that they echo longer around the body after the last spurt.

These changes give useful clues to the biological age of the woman's ovaries and her chances of conceiving naturally or by IVF. A blood sample, taken from her arm on the third day of her period is used to measure the levels of FSH. If they are more than twice as high as the standard in young women, ovarian age is already advanced and her prospects of getting pregnant are less, although not necessarily vanquished. It would make sense to use the ovarian hormones rather than pituitary FSH as direct indicators, although in practice they have not turned out to be better indicators of ovarian age. A low level of inhibin or an early rise in estrogen are equally bad omens for conception and pregnancy and are sometimes followed up with a pituitary "challenge test" to confirm the bad news or to provide some reassurance.

But none of these tests offer the certainties a woman worried about her biological clock needs. She may be puzzled that although we can screen for minute changes in the genes of a tiny embryo, we cannot tell her how many eggs she has left in her ovaries or whether they are still good for pregnancy. She may feel reassured about her chances while her periods last. But, although their existence is necessary, periods are not enough to indicate fertility and may come to a complete halt unexpectedly.

Methuselah's Caviar

The female body appears to be made of more robust stuff than the male's and it lasts longer. But in humans, the ovary seems out of time with the rest of the body and ages faster than any other organ. Unlike the testes, which turn out millions of sperm day after day into ripe old age, the ovary is endowed at birth with a limited store of eggs that is programmed to run out in middle age. Nature has carried out an act of biological sabotage on women. We do not know why, but we do now have a better understanding of how it happens.

A baby girl is endowed with a million tiny eggs in her two ovaries. Each egg rests in a nest of cells called a follicle. By the time the girl has reached her teens, the follicles have started to ripen for ovulation and to secrete the estrogen needed for conception and the support of pregnancy. When her ovaries have started to function and her periods begin, she has already lost three-quarters of her eggs, and worse is yet to come if the brain or any other vital organ lost cells at the same rate we would be in a sorry state by middle age! The disappearance of eggs involves a mysterious waste process called atresia in which follicles are absorbed into ovarian tissue, never to be seen again. At the age of 25, a woman is losing nearly 40 eggs a day and yet ovulates only one per month. If the ovary was a bank and the eggs the currency, this would be an extremely prodigal use of her capital.

Atresia goes on without respite throughout life, setting the course for menopause. By the age of 35, when many women are starting to think about having a family, egg numbers have fallen by 90% since their periods began. We might expect the ovary to adopt a more thrifty mode of operating to keep fertility options open for as long as possible. But surprisingly, the organ becomes even more wasteful than before and the eggs disappear twice as fast, running out around the age of 50.

Once the eggs have gone, there can be no further genetic investment in the future. This is not the way we would design a part to last in a Rolls-Royce, and it did not have to be this way for females either. If the ovary kept the stem cells for making new eggs after birth as the testicles keep theirs for making fresh sperm, menopause need never occur. Stem cells can make unlimited numbers of cells for as long as needed, and organs that retain them throughout life show fewer signs of aging, apart from the occasional mutation. The bone marrow produces 100 billion new red blood corpuscles per day and the testes a "mere" 100 million sperm daily. Their stem cells manage this feat by budding, but they ensure that for every new red cell or sperm cell exported another stays behind to keep the production line running.

Female frogs and bony fish keep stem cells in their ovaries to make new eggs and compete with the males for fecundity. They can drive their ovaries to maximum capacity and shed thousands or millions of eggs because they have to neither carry offspring nor take much responsibility, if any, for their larvae. A marine biologist once sent us a rather odd barrel of fish from a catch in the deep Pacific waters off British Columbia. The rockfish he netted are members of the perch family, whose main claim to fame is their extraordinary longevity, which, at 120 or more years, matches that of the freshwater sturgeon. When we examined the fish, we could not distinguish young ovaries from old ones. As far as we could tell, the eggs were identical, and it was a sight that the paragon of longevity, Methuselah, might have relished.

Evolution has molded a more mortal ovary in our species. The human female has lost the ability to replenish the organ every month with fresh eggs from stem cells. Since she draws from a finite bank of eggs, the decline in their quality as well as their quantity is inevitable. Eggs that are 40 years old are not as fresh as those that were made last month from a stem cell. A man's sperm at 40 is likely to be as good or as poor as it was ten years previously, and so it remains for the next two decades or longer, apart from an increased risk of rare mutations. In a menopausal ovary only a residue of eggs remains, and these old eggs are the least fertile. They are like the popcorn kernels left at the bottom of the pan after the rest— white, fluffy, and delicious—have burst. A woman's eggs begin to deteriorate during her thirties. This is the major reason an older woman has to wait longer on average to conceive and why the pregnancy is more likely to end in miscarriage.

If there were any thoughts that the problems lie in the uterus rather than in the eggs, they have been dispelled since IVF and egg donation. For women aged 30, 20% or more can expect to take home a baby after treatment in a good clinic, but for women in their mid-forties there is only a 3% chance. It has been calculated that at this rate, an American couple in this age bracket would have to lay out at least $120,000 to have a baby using the woman's own eggs. Transferring more embryos to increase the chances of having at least one that is healthy helps to buck the trend. But patients who have poor results may have to opt for egg donation from a woman under 35. Such a transfer would bring the woman's chances of a successful pregnancy in line with those quoted for the age of the donor. The age of the uterus seems to be irrelevant, for provided it remains healthy, this organ can perform as well at 50 as it did in the teenage years.

A chief cause of poor egg quality is chromosome abnormality, which often occurs when the egg is in the process of its meiosis division before ovulation. The great majority of the bad eggs are either never fertilized or fail to develop very far after fertilization. But occasionally a strong one carrying an extra copy of chromosome 21 manages to thrive in pregnancy and reaches full term as a baby with Down's syndrome. Numerically, most of these babies are born to young mothers because they bear most of the pregnancies, but the risk increases dramatically with age. Starting at a frequency of less than 1 affected fetus in 2,000 pregnancies for women in their twenties, the risks rise to a peak of about one in 50 for mothers in their mid-forties, although most are detected nowadays during prenatal screening can be terminated, if necessary. Even so, the risks should not be exaggerated or cause undue anxiety, because the great majority of babies born to older mothers are healthy and normal.

Chromosome errors often occur when the dividing egg fails to segregate the 46 chromosomes into two equal sets—one into the egg and the other into the polar body, which is discarded as a cytoplasmic bleb. When an extra chromosome is present—and there are then a total of 24 in the egg at fertilization—a normal sperm will create an embryo—and possibly an infant—with 47 chromosomes. If the egg is missing a chromosome, then the embryo will end up with only 45 chromosomes. This deficiency causes the embryo to fail to survive even the early stages of pregnancy, the only exception being when the missing chromosome is the X. Then a small number of affected embryos survive, and girls are sterile and short in stature (Turner's syndrome).

We are still surprisingly ignorant about why chromosome abnormalities occur more commonly in older mothers. According to one theory, more Down's babies are born to older mothers because the aging uterus relaxes its guard against abnormalities and more often fails to eliminate them. But today most scientists believe that the problem is due to the production of more abnormal eggs. It seems careless of the ovary to release a bad egg when the uterus is ready to make a last-ditch attempt at a successful pregnancy, although there may be an evolutionary logic. This is known as the Prisoner's Dilemma, and it is used to explain cooperative or altruistic behavior in nature.

Imagine two partners in crime who face a capital charge, like the notorious body snatchers Burke and Hare in Edinburgh in the 1820s. They may guess that the other's ploy is to keep quiet and they may then both get off with a light sentence. However, there is the risk that the other will confess and put the greater blame on his partner. This is exactly what Hare did, and, by turning Queen's evidence, got off scot-free and sent Burke to the gallows.

If I may anthropomorphize the chromosomes for the sake of an evolutionary argument, it is in the interests of them all to get into the egg. Otherwise they will perish and their genes will be lost. In a young woman, it makes sense for those in both sets to cooperate and leave the outcome to chance. If either set A or set B does not win in making the egg in this round, then it will perhaps be luckier in the next, when the dice are thrown again at ovulation. Cooperation is better than competition, because if they both get into the egg, neither could win; the embryo would either perish or it would produce a disabled child. But when the woman gets older and the time of menopause approaches, the stakes are higher because there may not be another chance. Therefore it pays A and B to compete and stay in the egg, even at the risk of disaster. This theory neatly explains the steeply increasing risk of chromosome abnormalities as women get into their forties, even if we have no idea how a chromosome could have a sense of time or urgency. It also fits in with the lack of such a marked age effect in men and laboratory animals, where menopause is absent.

The theory also suggests that younger individuals with fewer eggs and a potentially premature menopause are at greater risk of having a child with Down's syndrome, although we do not know if this is true in practice and have little idea how to avoid the danger. Perhaps some menstrual cycles are more hazardous for pregnancy than others, depending on the levels of the hormones. The only thing we know for certain is that even if a woman has regular periods and takes every dietary and health precaution, eggs do not improve with age like a good red wine. A woman could choose donor eggs instead, and this is the only way a postmenopausal woman can be helped to become pregnant. But few women who have yet to reach the change are so casual about their own genes. Most want to make strenuous efforts to have a baby that is theirs in every sense.

Researchers are trying to improve their chances by selecting the best eggs and embryos or by attempting to rejuvenate the eggs. The drawback in either case is that couples need to have an IVF procedure. Genetic diagnosis can be used to screen eggs or embryos for chromosome number so that only those with normal sets are used, and this may one day become standard IVF treatment for women over 35.

Some hopeful scientists are injecting cytoplasm from young eggs into old ones with the aim of invigorating them with more youthful constituents. Several babies have been conceived after cytoplasmic transfer, but although this is good news for the families concerned, the method is not yet scientifically proven. We are unsure whether eggs from older mothers are poorer in quality because of deterioration in the nucleus or the cytoplasm, or both. An alternative strategy would be to transfer an aging nucleus into young cytoplasm. Perhaps the best way to test this would be to dissect embryos at the four-cell stage into separate cells and fuse each of them with an enucleated egg from a younger donor. They could then be transferred to the uterus in pairs—not necessarily from the same embryo if more are available, to avoid the risk of creating genetically identical twins. The spares would be kept as backups in the freezer. This is, of course, a form of cloning, but it is more acceptable than cloning with somatic cells as both the mother's and the father's sets of genes would be present, and the prospects of a healthy outcome are better.

Until we understand more about the biology of eggs and can protect them from aging, we can expect to see these and even more desperate measures being contemplated. It must surely be better to preserve and protect an egg than to tinker with an embryo. But while the search goes on to improve the success of ART technologies, there are other solutions in the pipeline. One is even at hand.

Banking on the Future

Anyone expressing a belief that aging will soon be conquered is likely to raise some eyebrows and be considered rather naive. Old age is the final and inevitable act in life's play, and menopause marks the opening of the last act. Or so it used to be regarded. Happily, such defeatist and pessimistic views are expressed less often now that people so often enjoy a prolonged phase of healthy and active life after the childbearing years have passed. Since the advent of egg donation, it is possible to have another child after menopause, though it would be preferable to find a way of preserving a woman's own eggs, either by freeze-banking or even slowing the aging process in the ovaries. Most of the eggs in the ovary, at least at younger ages, are probably normal, yet they are still wasted. If even a small fraction of them could be saved, menopause could be staved off for a while so that fertility is improved in the thirties and forties when it counts most. Even if the general aging process is inexorable and only the accelerated loss of follicles in the last 15 years of menstrual life can be prevented, menopause could be postponed from age 50 to age 70.

The most attractive method would be to take an ovarian preservation pill. If the production line in the ovary could be halted for a few years, the eggs that otherwise go to waste might be saved to be used later. When a woman takes the oral contraceptive pill to suppress the pituitary hormones FSH and LH, the monthly ovulation is prevented. But halting the ovulation of just one egg per month has no significant effect on the declining numbers of eggs overall.

Like women who suppress ovulation naturally by repeated pregnancies, the timing of menopause and aging of the ovaries still occur on schedule, usually between 45 and 54 years of age. This is not the pill to stop ovarian aging, though there may be one in sight.

The search is on for a way of stopping the atresia of the immature eggs in the ovaries. If a way were found to prevent them growing, a "career" pill might be designed to inhibit fertility for as long as required. Whether we would end up with better eggs or just more of poor quality is a moot point that will need to be tested experimentally. If all were well, the pill might be taken for some years as a young adult, extending the number of menstrual years accordingly. But shutting down the production of estrogen from the ovaries would trigger menopausal flushes and the onset of bone loss and vascular problems. The pill would have to contain some estrogen as hormone replacement to protect women from these hazards. It may sound strange to take HRT before menopause, but in fact the estrogen in the standard birth control pill acts as a form of HRT, while the progesterone prevents ovulation. The proposal is perhaps not then so bizarre as it sounds. Besides, the lower levels of estrogen in HRT than during a natural cycle should actually reduce breast disease. The timing of their true menopause could then become a matter of choice, and they could always change their minds and have their ovaries removed at any time to bring on a natural postmenopausal condition.

Designing a pill to preserve the ovaries and protect against unwanted pregnancy at the same time may be a goal for the future, but frozen egg banking has already arrived. If a biopsy from an ovary or better still the whole organ is deep-frozen to liquid nitrogen temperatures (nearly minus 200 degrees Celsius), the tissue would probably be fresh and vigorous when it was thawed years or decades later. So successful has this method been that some cancer patients are already storing tissue in the hope of conceiving after they have made a good recovery. Healthy women who want to retain the choice of a late baby may also wish one day to avail themselves of the opportunity.

The story of frozen ovary banking began 50 years ago in London when a group of scientists tried to revive sperm after deep-freezing. There were many failed attempts with different chemicals to protect the cells from ice crystal damage. One day a thawed sample was found still to contain wriggling sperm with almost 100% in the sample alive. For several days the scientists were baffled by a result they could not repeat until suddenly they realized that there had been a mix-up in the fridge with another bottle that had lost its label. The secret of success was found to be the natural chemical glycerol, and this was the first effective "cryoprotectant." Later it was discovered to occur in some insects that protect themselves from low temperatures with this natural antifreeze.

Having discovered the benefits of glycerol, the team pioneered sperm banking and revolutionized artificial insemination for the beef industry. They also showed that rats' ovaries could be stored deep-frozen in dry ice, although it was not until recently that the methods were perfected with the use of other cryoprotectants and automated freezers to control the cooling rate. This aside, there had been virtually no further progress for 30 years. Then, using sheep from the flock that produced Dolly, my colleagues and I managed to freeze the ovaries of these animals and return them as transplants several weeks later. All the ewes started cycling again, and when they came into sexual heat, some of them were mated. Five months later they delivered healthy lambs. These have been heady days for low-temperature biology, and the similarity between sheep and human ovaries bodes well for practical applications.

In the past, cancer specialists were concerned mainly to keep their patients alive for as long and as comfortably as possible. The disease could be warded off only for a while, so the fact that the drugs and radiation treatment caused sterilization and menopause was of little moment. Patients these days often receive more aggressive treatment but are far luckier. A number of former killer diseases of young people, such as Hodgkin's disease, can now be cured. Once they have recovered, patients naturally want to resume a normal life, including marriage and a family. Ovarian banking has been developed to enable any patients who receive sterilizing chemotherapy to store eggs for their own later use.

The freeze-banking of ovarian tissue could be useful in a large number of conditions. Tissue is often removed during surgery for an ovarian cyst or endometriosis, and it could be stored "just in case." Other women, with a mutation exposing them to potentially fatal ovarian cancer, might want to have their ovaries removed prophylactically before the disease sets in. They would be wise to use HRT to avoid the problems of early menopause, and they would be prudent to store some of their tissue if they have not completed their family or to keep open the possibility of having future children.

The storage of ovaries, or at least biopsy specimens taken from them, currently looks more promising than the freezing of mature eggs, which has a low success rate and also carries the costs and trouble of an IVF procedure. Even a small slice of ovary contains thousands of immature eggs compared to the ten or so recovered after hormonal superovulation for IVF. Tissue storage has turned out to be relatively easy. After soaking and cooling in a cryoprotectant solution, specimens are transferred to a steaming tank of liquid nitrogen, where they can be left for years without significant deterioration. There are currently no regulations that set a "sell-by" date on freeze-banked ovarian tissue, as there are in the case of embryos and sperm in some countries. There is cause for concern about the long-term security of the specimens. In the case of child patients, for example, tissue may not be required until long after the staff who froze it have moved on or retired, and regulations may be required to safeguard the patients' interests.

Until someone becomes pregnant from using their stored tissue it is too soon to celebrate, but we have more confidence in the technique because there is more than one option available to help to have a baby. The most attractive strategy is to transplant tissue back into the same patient to restore natural fertility, as we did in the sheep experiments. Whether this is wise in the case of cancer patients depends on whether any malignant cells might be lingering in the graft. A possible future use is for "egg farming," which, as we saw in Chapter 5, will be needed if human cloning is ever to be practiced on a large scale. Eggs would be extracted from thawed tissue and, still in their nest of follicle cells, ripened in petri dishes until they were ready to be fertilized in vitro. The embryos would then be transferred to the womb after the woman had been treated with estrogen and progesterone to mimic the ovarian hormones she lacked and to prepare for implantation. Egg farming avoids any risks from ovarian transplants, but it is still an experimental procedure under development.

Last, the eggs could be grown in animals that have been rendered immunologically deficient by drug treatment or a mutation. This trans-species graft—or "xenograft"—is justifiable only in extremis when there is no other way of producing an egg without exposing a woman to potentially malignant tissue. We hope that the development of egg farming will obviate the need ever to use xenografts. If we did use them the animal would not, of course, become pregnant as the egg would be removed from it and fertilized with a human sperm in vitro and then transferred to a human womb. Such a method may work, but I would not recommend growing eggs in animals until a great deal of research has been done to reassure us that there are no risks from viruses or prions jumping species and infecting the egg and affecting the baby.

The methods that are being developed to preserve fertility for sick people could equally be used for the healthy. Storage of ovarian tissue is already being offered by private clinics for a fee. Before long we shall probably see many more women electing to have healthy tissue banked to keep their eggs "green" rather than conceiving with an old egg prone to defects. Whatever our attitude to the technology of postponing fertility, it must surely be a better principle to store eggs rather than embryos.

With the storage of healthy tissue for personal use there are no concerns about transmitting disease, so reimplantation of tissue is the natural option. Looking ahead to a time when the problem of organ rejection is solved without the need for risky immunosuppressive drugs, it is possible to imagine ovarian tissue donation becoming routine. Tissue would be advertised for sale on the Internet, just as donor eggs and sperm are now. If egg donation for IVF treatment is so widely accepted, why not tissue donation?

The technologies under development will liberate reproduction from its natural time constraints. A woman will be able to keep her best eggs for as long as she wants and try for a pregnancy when she chooses. And once tissue is reimplanted in the owner's body, there will be none of the legal restrictions on fertility that currently restrict IVF treatment in many countries. A woman of any age, perhaps even carrying donor tissue from someone with enviable genetic qualities, will then be able to have as many babies as she wants and at whatever age she chooses. What is more, the graft would give her a monthly reassurance that it was still working. If these methods of postponing menopause and deferring the childbearing years materialize we can expect to hear plenty of adverse comment and censure.

Conquering the loss of healthy eggs in the ovaries is said to subvert a natural process, yet we already interfere with aging in so many ways. We use dietary control and insulin to offset the increasing glucose intolerance of aging and we give HRT for postmenopausal problems without concerning ourselves much about the use of such artificial remedies for natural conditions. Doctors look both ways. On the one hand, they see all the health problems that menopause heralds, and on the other they are grateful for being relieved of the responsibility for obstetric complications in older women. Biologists have ambivalent views, too. Some regard menopause as having no purpose, because it occurs after the stage when most people have completed their families and long after our hominid forerunners would have survived. Others regard it as having had a useful place in the past. According to the "grandmother theory," it is in the interests of a woman to give up having more children after a certain age because her genetic line is better served by helping to raise her grandchildren than by negotiating a risky pregnancy. In other words, grannies have always mattered.

Reproductive technology has to plot a course between these conflicting currents and, hopefully, will offer women more choices in the future. People are now choosing the number and timing of births with more thought than ever before. This is very welcome. It is in their interests to produce a healthy child while they have enough years ahead to raise him or her successfully. ART helps them make a decision at their discretion, and the rest of us should trust them to design the best possible future for their babies. In years to come, I suspect that we will look at what has happened to the children of these older parents and wonder what all the fuss was about.

9

Reproductive Liberties

A beautiful actress once playfully proposed to the Irish playwright George Bernard Shaw: "We could produce remarkable children together." "Ah yes," he replied in a flash, "but what if they had my looks and your brains!" We may have come a long way in science and technology, but the reproductive process still turns out far from predictable products. Beyond the basic desire for a healthy child free of any physical or intellectual defect, parents usually have plenty of other preferences besides. They may even dream of a technological Apollo or Artemis—perfect in beauty, reason, and morality. Perfection is, of course, illusory, and the goddess Artemis was always associated with the moon—a classical symbol of the unattainable.

Many people fear the extraordinary progress of modern reproductive technology in our consumer-oriented age. They fear that would-be parents, in making choices about the kind of child they would like to have, will think of the child-to-be more as a fashion object—the designer baby—than a unique human being with its own needs. Yet, the decisions that prospective parents make about when to start a family, whether to have gamete donation, and genetic testing and attitudes to medical termination are so personal that any generalization is futile. What is more, although the great breakthroughs in reproductive science are recent, the notion itself of the designer baby is not new. People have always sought to influence the characteristics of their children. Their concerns in this respect not only reflect the size of the investment they are making—the cost in time, money, and emotion—but also a natural desire to have the very best for their child. If an older couple had only six eggs left from which to make a baby and they knew that three of them were blemished, it would be perverse to mix the eggs up and choose one at random for fertilization. This would be counterintuitive and run against the whole tradition of science and medicine, which strives to avoid and overcome disease and disability. Also, more people are now better informed and approach fertility with greater seriousness than ever before. Focusing on a few precious children encourages a sense of responsibility and careful planning of the family-to-be. If the parents restrict their number and timing, the children should benefit in almost every way. But in spite of the many advances, the role of science is still a limited one. Even more crucial to a child's future—and more difficult to predict the consequences—will be the social and educational opportunities the child enjoys in the course of his or her life.

The suggestion of improving the quality of the reproductive process quickly provokes the old specter of eugenics. It is undeniable that favoring a healthier type of gamete, embryo, or fetus is eugenic in principle, but it is a far cry from coercive policies that have been practiced in the past. The great difference is that nowadays, the most important decisions are made by the couple concerned in consultation with their doctors and counselors. Even if we could use science to determine superficial characteristics—say, hair color—need we worry if the decision did not disadvantage the child in any way? As a concern it pales into insignificance in comparison to the real moral dilemmas that abortion poses. In the United Kingdom, there were 180,000 abortions in 1997, only 2,000 of them on grounds of abnormality. It would be incongruous to condemn actions based on personal preferences if we accept the destruction of so many normal fetuses.

In the future there should be few occasions when a seriously deficient baby is born without warning. The fear is that the additional information we gain about an embryo from screening its genome will encourage the desire for a more standardized product. But the critics often have more faith in the ability of science to predict and manipulate than have the scientists themselves. Sensationalist media coverage encourages a misplaced belief in science's omnipotence, although the revolutionary pace of discovery is real enough and it is certainly a remarkable time to be a biologist. In the past decade alone we have seen break-throughs in ICSI, PGD, and cloning, as well as many other important discoveries. In theory, anything is possible—even completely redesigning the human body—provided we do not contravene the bedrock laws of physics. In practice, successful scientists tackle projects that seem to be most tractable to them and likely to be for the public good. The main issues on their current agenda are the characteristics of normal and harmful genes, problems of infertility, and, a little further off, perhaps, the practical implementation of cloning and germ line therapy. These are weighty enough subjects, so why worry about really monstrous developments that may never happen? As one of Benjamin Disraeli's dictums reminds us, "What we anticipate seldom occurs; what we least expect generally happens." The responsibility of scientists is to warn when they apprehend new developments that have social significance, and the media's is to accurately report facts in the interests of the public.

Some of the traits we value most, such as intelligence, are still beyond the ability of science to manipulate. Even when we have a better idea of what to test for—and in the case of IQ that will include a large number of genes—the correlation between the genotype and the phenotype is likely to be fuzzy. As for improving on nature, any technique involving the brain will need to be hugely sophisticated to improve on the original model. Genes never act in isolation from each other—they interact with each other and are affected by external influences. Such interaction means that tinkering with one trait is likely to affect another in some unforeseen way. There is in any case a natural trade-off in the way genes interrelate, one of the best examples of which is the aging process. Over the course of evolution, nature favored genes that produced beneficial effects at young ages. Even if a price had to be paid in degenerative problems later on, it didn't matter because so few people lived long enough to be affected. But if now, for the benefit of a longer-lived population, we should seek to favor the late effects of some genes, we might end up with poorer health—and less fertility. In addition to these uncertainties, there are the vagaries of the environment and the random factors in the nature of life that can never be fully controlled. While the struggle against the mistakes of nature will last as long as science itself, no gamete or embryo will ever be absolutely safe, nor can any clone be perfect.

Leaving things to nature can expose us to hangover effects from evolution that makes us ill-prepared for an artificial world. Genes that are good in one environment can prove to be less than ideal in another. There is much about human physiology that is ill suited to modern life, including such familiar problems as fever, depression, and morning sickness—which, according to the emerging field of "evolutionary medicine," served useful purposes in the past. Let us take one example. In the Micronesian island atoll of Nauru, over 60% of the adult population has diabetes and obesity since it adopted a Western diet and life style, and many other Asian populations have similar problems. Some genetic variations that affect metabolism and were helpful in the days when food supplies were unreliable and famines were common are unsuited to times of plenty. There is certainly no case for genetic screening or intervention where a carefully controlled diet and health education are available today, but harder decisions may be ahead for all of us. If the earth became contaminated by pollutants or levels of solar radiation soared, it could become desirable to acquire genes conferring more resistance to mutagenic chemicals and harmful rays. Since natural selection operates so slowly, we would have to turn to technology if we wished to protect future generations.

At present, we are able to screen embryos and fetuses for abnormalities but cannot rectify any faults that we might find. Progress is being driven by advanced scanning methods for visualizing the baby's developing form in utero and by testing for harmful mutations at even earlier stages. As the Human Genome Project reaches its conclusion and automated DNA sequencing becomes more routine, the genes responsible for our physical and behavioral features are being identified and even the rarest anomaly becomes detectable. Technology tends to be expensive at first but then becomes cheaper and widely available, just as television and computers are now affordable for virtually all sectors of our society. Eventually, there could even be DIY kits for genetic testing, like pregnancy and paternity tests, which would short-circuit the delivery of genetic counseling services. The question that is now being asked is whether this is leading to a narrower definition of a fine baby.

There is a concern that as certain types of abnormality become rarer because of the ability to pick and choose, attitudes could harden towards the few babies still with problems. Prejudice against disability and ugliness is a problem that goes back centuries: Caliban, just like his real-life counterparts in Shakespeare's day, was detested and accused of being born of the devil. But many people who know individuals born with a physical disadvantage or a learning problem understand the emptiness of stigmatization and how much can be learned from the individuals concerned. Kylie, a little girl I know who was born with Turner's syndrome, says that when she grows up she wants to look after Down's syndrome children. This reflects a mature sensitivity to other disadvantaged people whom she perceives to be more needy than herself. Only by being aware of such people and bringing them fully into the community can we provide the best circumstances for them to lead a full and accepted life.

As soon as gene therapy becomes truly effective in adults and children, pressures will grow to use it in rescue missions before birth. This endeavour is likely to start with fetuses affected by the more common ailments, such as cystic fibrosis, and then move steadily to embryos at earlier stages of pregnancy and perhaps even to eggs or sperm. The attraction of this germ line therapy is that substitution of defective genes with healthy ones will not only overcome the fault in particular embryos but purge those mutations from families who may have carried a disease "in the blood" for generations. But the difficulties of the technology are sometimes underestimated and the likelihood of widespread application has been exaggerated. Egg and embryo therapy will depend on the readiness to intervene around the time of conception, which is not a trivial procedure for a woman. Only a tiny minority of people are currently using IVF treatment and, unless the procedure is simplified or the benefits are perceived to be great, germ line therapy is unlikely to be practised on a large scale for some time. For the foreseeable future, most people will continue to reproduce like the savages in *Brave New World*—leaving the mix of genes to nature and chance, and checking the health of the products during pregnancy. History teaches us that science has a disturbing habit of hitting on completely unexpected possibilities, and, if we ever manage to collect embryos efficiently from the womb after natural conceptions and replace them, I will have to eat my words. This remarkable step would allow us to routinely check the quality of embryos at the most tender stage, and many couples would then be tempted to try to influence the destiny of their child.

One of the fears most commonly expressed—and most difficult to allay—is that an enterprise that begins with the laudable aim of avoiding birth defects or disease and suffering later in life ends up as a program to engineer a "perfect" child. The ability to enhance human appearance and performance would be regarded as a zenith for technology but a nadir for society. This reaction is understandable, but we should remember that we already try to enhance the advantages of our children in every other way and people use artificial aids to increase their own creativity and achievement. Coleridge wrote much of "Kubla Khan" in a haze of opium, Aldous Huxley wrote *The Doors of Perception* under the influence of mescaline, and athletes have used performance-enhancing drugs to break records. These cases may be extreme, but there is no shortage of more innocent examples. Thousands of people are experimenting with

Viagra to improve their sex lives, or doping their scalps with Rogaine to encourage their hair to regrow. In both cases the problems are due to the natural process of aging, and the remedies are themselves enhancing drugs.

Faith in an idealized natural state is deeply rooted whereas technology which interferes with nature is sometimes portrayed as alien or sinister. There are, of course, memorable instances in which an invention has proved disastrous, but something should not be condemned simply because it is "unnatural," for that disposition would be completely at odds with the lives we lead. The anthropologist Margaret Mead, who lived among South Sea Islanders and New Guinea tribes, pointed out that to describe traditional life styles as natural and Western life as unnatural is sentimental nonsense since all societies use artifacts to meet their particular needs and circumstances. This observation has perhaps even greater force today when our environment is so palpably changing, whether due to human negligence or some cosmic cycle. As with every other species, the evolutionary imperative is to adapt or perish, and biomedicine may have to be even more inventive to protect future generations. There is a paradox here. While the conditions of natural conception are said to be deteriorating, possibly from changes in human habits and diet, conception in the test tube is improving all the time from the benefits of new knowledge.

I started this book with the story of Victor Frankenstein because he dared to create life. Motivated by self-interest, he made a monster who was miserable and unloved. In the pioneering days of ART, it was said that conception in test-tubes would produce monsters too, but, instead, thousands of homes now ring with the voices of happy families and there are no more birth defects than usual. The main problem has been in producing too many multiple pregnancies, which are neither in the interests of the children nor their parents. Despite the celebratory mood, we must remain watchful as children conceived with an ever expanding array of ART reach adulthood. New discoveries which promise much for the future, such as germ line therapy, should be regarded warily, and any attempts to experiment recklessly with the life of a child-to-be should be resisted. The interests of the children concerned should always be paramount when a new life is planned. This principle is enshrined in the Human Fertilization and Embryology Act (1990): "A woman shall not be provided with treatment services unless account has been taken of the welfare of any child who may be born as a result of the treatment."

The urge to reproduce is so great that new techniques are sometimes tested in clinics with daring haste, before full assurance about the long-term implications can be given. Many scientists are worried about the gulf that is growing between our technical wizardry and understanding of the biological implications. These are compelling reasons to encourage research to fill some of the more glaring gaps in our knowledge lest free enterprise unleashes untested technologies on a ready market. Regrettably, federal funding in key areas is forbidden in the United States and the European Code of Bioethics has also pronounced against studies on human embryos. The price of such restrictions manifests itself not just in a slower pace of progress in treating infertile couples but in our continuing ignorance about the crucial formative days when life-long handicaps often arise.

The march of science will continue despite obstructions in its path and the demonization of some of its leaders. I do not think we will see scientists renouncing their powers like Shakespeare's master magician Prospero, who, in *The Tempest,* broke his staff and destroyed his book of spells. Fascination with the process of human development and the dream of overcoming disease drives them on. However, we have to pause from time to time to seek safety assurances before progressing, as molecular biologists did in the 1970s when DNA technology was in its infancy. I hope that in the future the technology of creating new life will become less complex and alienating and less reliant on experts who currently make all the key decisions. One of the benefits of new knowledge could be that, instead of using highly invasive methods, we might be able to gently work with natural processes. The pioneer of test-tube conception, Robert Edwards, has been calling for refinements in IVF practice that will make treatment simpler, less expensive, and almost completely safe. Likewise, Edward de Bono, principal exponent of lateral thinking, has recently emphasised simplicity as a touchstone for the social acceptance of future technology. Just as computer interfaces are becoming more user-friendly, and consequently universally accessible, we should be optimistic for ART, even though specialists will still be needed to deliver the services.

In many countries, though less so in the United States, rafts of laws have been introduced to control and restrict ART. However, where human desire is strong prohibition is often futile, as we have seen from the banning of alcohol in America in the 1920s and with cannabis today. If it is difficult to control recreational substances, how much harder will it be to control the powerful urge to produce a child? People already make a nonsense of regulations by traveling to countries where there is greater liberty to use a technology that can help them. In Germany laws regulating IVF technologies are among the most restrictive in the world, but a young couple can board a train in Düsseldorf and quickly reach Brussels or London for treatment that is forbidden at home.

Britain, which is credited with some liberality in ART as well as a large share of the major scientific breakthroughs, still imposes strict licensing of fertility treatment. Yet it seems unfair to legislate when other people, not so constrained by infertility, can act freely. Most natural conceptions occur in circumstances that are far from optimal, and many are casual: a recent survey in America revealed that 56% of pregnancies were unplanned. Surely when couples make careful plans for conception using ART they should not be discriminated against simply because they do not have as much freedom to choose when, how, or with whom they make a baby. The famous case of the Nottinghamshire woman Diane Blood reveals how heartless legal proscriptions can be. After the sudden death of her husband from meningitis, she was refused permission to be inseminated with his sperm, which had been collected after he had lost consciousness. The legal obstacle was that he had not left written consent. Despite almost universal expressions of sympathy she had to go to the Court of Appeal to have the decision overturned. Even then she had to be treated by AI overseas, where it has proved successful. Such examples warn us to go gently when drafting laws. Legislators—usually a generation older than hopeful young couples—should bear in mind the potential misery caused by denying life's most precious gift.

Mrs. Blood declared: "The state should only interfere in the lives of individuals and couples when absolutely necessary." Many others share her concern about the dangers of overbearing authority in the private matter of reproduction, and no one has expressed this more forcefully than Ronald Dworkin, a professor of jurisprudence at New York and Oxford. Writing in the context of abortion and euthanasia, he argues that "procreative autonomy" is a natural corollary of amendments of the American Constitution that guarantee freedom of religion and choice in a way of life according to personal beliefs. He defines this autonomy as "a right to control their own procreation unless the state has a compelling reason for denying them that control."

Public trust in government has reached a low ebb and the old fear of eugenics still lurks at the back of our minds. Citizens now feel safer looking after their own interests. Dworkin's British counterpart, John Harris, argues that free parental choice is a good defense against bias on the grounds of race, gender, or disability. Perhaps it is better to accept a minority of distasteful personal preferences in the reproductive marketplace than to legislate heavily. It would be far more alarming if the state, as guardian of the public purse, chose to make reproductive decisions for us by deciding the kind of people that there ought to be, and weighing the costs of raising a defective child. That would be the slippery slope to social eugenics, and there is no scientific justification for revisiting it. No tests have yet been devised to define who would make a good parent or what kind of life is worth living.

Perhaps a more libertarian stance is the best safeguard against the emergence of a Huxleyan scenario, although there is still a danger that social harmony could be lost from the selfish use of reproductive technology. Reproduction is a private affair and parenthood is its intense reward. The reproductive drive needs to be tempered by the collective virtue of human solidarity so the aspirations and rewards of having fine children can be shared by everyone. When Prospero's daughter Miranda exclaimed those famous words—"brave new world"—it was with joy and wonder at the appearance of previously unknown human creatures on her desert island. She also dreamt of a richer and riskier world beyond her limited horizon. Leaving the island meant a loss of security and coping with unfamiliar surroundings, as all voyages of discovery have to pass through uncharted waters. The enterprise that we have embarked on in reproductive science also has much more to welcome than to fear.

Glossary

amniocentesis Medical procedure involving the withdrawal of a sample of the **amniotic fluid** that surrounds the fetus for genetic analysis of cells sloughed from its body surfaces.

amniotic fluid Fluid in which the fetus floats (as well as swallows and excretes into) throughout pregnancy.

artificial insemination (AI) Injection into a woman of sperm derived from either her male partner or a donor.

ART Assisted reproductive technology, including **IVF**.

biopsy Small fragment of tissue or a few cells removed for diagnostic purposes.

chorionic villus sampling (CVS) Removal of a tissue **biopsy** from the fetal membranes for prenatal diagnosis of disease.

chromosome Paired, thread-like structure in the cell nucleus containing a core of DNA; there are 46 chromosomes in normal human cells, including either two X (female) chromosomes or an X and a Y (male).

clone Genetically identical copy of a molecule, a cell, or an organism.

cryoprotectant Substance used to protect cells or tissues from damage during freeze-storage.

cytoplasm Cell contents excluding the nucleus.

DNA (deoxyribonucleic acid) Very long molecule at the core of each **chromosome**, consisting of bases (abbreviated A, C, G, T), sugar, and phosphate, that linearly encodes the genetic instructions of a cell and is copied every time a cell divides.

ectogenesis Growth and development of an embryo and fetus outside the body (i.e., in an "artificial womb").

embryo Earliest stages of development—by convention from fertilization to approximately six weeks of development when the organs are forming.

embryo transfer   Procedure in which embryos created by **IVF** are transferred in a fine tube to the uterus for pregnancy.

enzyme   Molecule that catalyzes a specific chemical reaction in the body.

estrogen   Steroid hormone produced by the ovary and placenta that promotes growth of female sexual characteristics (e.g., breasts) and prepares the **uterus** for pregnancy.

eugenics   Ideology proposing selective breeding to avoid defective fetuses and improve genetic stock.

fallopian tubes   Paired tubes for conveying sperm to the site of fertilization near the ovaries and for transporting embryos in the opposite direction for implantation in the **uterus**.

follicle   Cystlike structure in the ovary, each of which nurtures a single egg or **oocyte**.

FSH (follicle stimulating hormone)   Hormone secreted by the **pituitary gland** that stimulates **follicles** to grow and secrete **estrogen**.

gamete   Male or female reproductive cell (sperm or egg).

gene   Unit of inheritance, represented by a specific sequence in the **DNA** molecule that encodes instructions for making a specific protein through an intermediate **RNA** molecule.

gene, dominant   Variant ("allele") of a gene that always manifests itself regardless of the other gene copy on the complementary chromosome.

gene, recessive   Variant ("allele") of a gene that is not manifest when a **dominant gene** is present.

gene therapy   Treatment of a disease by rectifying error(s) in the **DNA**.

genome   Entire "library" of genes in an individual.

genotype   Particular genetic constitution of an individual.

germ cell   Primitive precursor cell of sperms and eggs.

germ line therapy   Correction of a genetic defect in the reproductive cells (gametes and embryos), producing heritable effects.

gonad Gamete-producing organ (ovary or testis).

HCG (human chorionic gonadotropin) Hormone that is chemically similar to pituitary **LH** but is secreted by the placenta; it is used in pregnancy tests and fertility treatment.

ICSI (intracytoplasmic sperm injection) Modification of the basic **IVF** technique used to assist couples with severe male infertility; a single sperm is injected into an egg to aid fertilization.

IVF (in vitro fertilization) Fertilization by mixing eggs and sperm in a laboratory test tube or petri dish.

laparoscopy "Keyhole" or minimal-access surgery using a miniature telescope to view internal organs of the body.

LH (luteinizing hormone) Hormone secreted by the **pituitary gland** that triggers **follicles** in the ovary to ovulate.

meiosis Special division that occurs only in gametes; it halves the number of chromosomes after DNA is exchanged between complementary pairs.

menopause Final menstrual period in a woman's life (usually around 50 years of age).

mitochondria Tiny sausage-shaped replicating structures that provide energy in cells.

mutation Error in the code of a gene that may be benign or affect cell function and cause disease.

nucleus Sac containing the **chromosomes** in the cell.

PCR (polymerase chain reaction) Method for amplifying short lengths of the **DNA** molecule for genetic analysis.

penetrance Degree to which a gene is expressed or its effects are manifested.

phenotype Characteristics of an individual, physical, physiological, and behavioral; compare **genotype**.

pituitary gland Pea-sized gland lying under the brain that secretes a number of hormones regulating growth, metabolism, and reproduction.

polymorphism Variant form of a gene or protein.

preimplantation genetic diagnosis (PGD) Genetic diagnosis by removing one or a few cells from an early embryo to test whether it is blighted with a harmful mutation or abnormal chromosome number.

prion Protein responsible for transmitting certain diseases, e.g., BSE and scrapie.

progesterone Steroid hormone from the ovary and placenta that is required for pregnancy.

reproduction Whole process of generation from production of gametes through fertilization, pregnancy, birth, and infancy to full maturity.

RNA (ribonucleic acid) Family of molecules, many of which carry an encoded message from the genes for making specific proteins in the cytoplasm.

somatic cells All cell types in the body except the **germ cells**.

somatic cell therapy Treatment of disease by correcting an error(s) in the genes of **somatic cells**; compare **germ line therapy**.

sperm Male reproductive cell or gamete.

surrogate Woman who carries a child on behalf of another woman.

transgenic Organism whose genome has been altered by the insertion of foreign DNA, usually with the intention of producing a different **phenotype**.

trophoblast Invasive cells that form the placenta.

twins, dizygotic (DZ) Twins formed from separate fertilized eggs and therefore genetically dissimilar.

twins, monozygotic (MZ) Twins formed by the splitting of an embryo into two and therefore genetically identical.

uterus The womb; pear-shaped muscular organ with a glandular lining in which the embryo implants and pregnancy is sustained.

yolk sac Transitory structure in mammalian pregnancy (but not marsupials) that serves as the first placenta; it is where the first blood cells are formed and where **germ cells** are formed.

zygote Fertilized egg before it starts dividing to form an embryo.

Further Reading

Chapter 1   The Myth and the Monster

Atwood, Margaret (1987). *The Handmaid's Tale.* Virago, London.

Bowling, A. (1996). Health care rationing—The public debate. *British Medical Journal* 312, 670–674.

Haldane, J. B. S. (1924). *Daedalus or Science and the Future.* Kegan Paul, Trench, Trubner, London.

Huxley, Aldous (1932). *Brave New World.* Chatto & Windus, London.

Jones, Steve (1993). *The Language of the Genes.* HarperCollins, London.

Nelkin, Dorothy (1995). *Selling Science: How the Press Covers Science and Technology.* W. H. Freeman, New York.

Nelkin, D., & Lindee, M. S. (1995). *The DNA Mystique: The Gene as a Cultural Icon.* W. H. Freeman, New York.

Rifkin, Jeremy (1998). *The Biotech Century: The Coming of Age of Genetic Commerce.* Victor Gollancz, London.

Rowland, Robyn (1992). *Living Laboratories—Women and Reproductive Technologies.* Indiana University Press, Bloomington.

Silver, Lee M. (1998). *Remaking Eden: Cloning and Beyond in a Brave New World.* Widenfeld & Nicolson, London.

Venter, J. C., et al. (1998). Shotgun sequencing of the human genome. *Science* 280, 1540–1542.

Vines, Gail (1995). Every child a perfect child? *New Scientist* 25, October, pp. 14–15.

Chapter 2   The Precious Child

Balen, A. H., & Jacobs, H. S. (1997). *Infertility in Practice.* Churchill Livingstone, Edinburgh, London, & New York.

Bowling, Ann (1996). Health care rationing: The public's debate. *British Medical Journal* 312, 670–674.

Caldwell, John C. (1982). *Theory of Fertility Decline.* Academic Press, London.

Dunbar, R. I. M. (1995). *Human Reproductive Decisions.* Macmillan, London.

Gosden, R. G., Dunbar, R., Haig, D., et al. (1998). Evolutionary interpretations of the diversity of reproductive health and disease. In Stearns, S. C. (ed.), *Evolution in Health and Disease*, pp. 108–120, Oxford University Press, Oxford, New York, & Tokyo (in press).

Grubb, W. Norton (1982). *Broken Promises*. Basic Books, New York.

Jansen, Robert (1997). *Overcoming Infertility*. W. H. Freeman, New York.

Packard, Vance (1983). *Our Endangered Children*. Little Brown, Boston.

Ober, W. B. (1984). Reuben's mandrakes: Infertility in the Bible. *International Journal of Gynecological Pathology* 3, 299–317.

Shoen, R., Kim, Y. J., Nathanson, C. A., et al. (1997). Why do Americans want children? *Population and Development Review* 23, 333–358.

Silver, Lee M. (1997). *Remaking Eden: Cloning and Beyond in a Brave New World*. Avon Books, New York.

Stone, Lawrence (1977). *The Family, Sex and Marriage in England 1500–1800*. Harper & Row, New York.

U.S. News & World Report (1998). The Cost of Children. March 30.

Zelizer, V. A. (1985). *Pricing the Priceless Child: The Changing Social Value of Children*. Basic Books, New York.

Chapter 3   The Pursuit of Perfection

Daly, M., & Wilson, M. (1978). *Sex, Evolution and Behavior*. Willard Grant, Boston.

Darwin, Charles (1871). *The Descent of Man and Selection in Relation to Sex*. Appleton, New York.

Dawkins, Richard (1989). *The Selfish Gene*. Oxford University Press, Oxford

Dickeman, M. (1979). Female infanticide, reproductive strategies and social stratification: A preliminary model. In Irons, W., and Chagnon, N. (eds.), *Evolutionary Biology and Human Social Behavior: An Anthropological Perspective*, pp. 321–368. Duxbury, North Scituate, Mass.

Dunbar, R. I. M. (ed.) (1995). *Human Reproductive Decisions*. Macmillan, London.

Duster, Troy (1990). *Backdoor to Eugenics*. Routledge, New York & London.

Furlow, F. B., et al. (1997). Fluctuating asymmetry and psychometric intelligence. *Proceedings of the Royal Society of London B* 264, 823–829.

Futuyma, Douglas J. (1986). *Evolutionary Biology*, Second Edition. Sinauer, Sunderland, Mass.

Galton, Francis (1979). *Hereditary Genius: An Inquiry into Its Laws and Consequences*. J. Friedman, London (originally published 1869).

Gould, Stephen Jay (1981). *The Mismeasure of Man*. W. W. Norton, New York.

Gould, Stephen Jay (1989). *Wonderful Life*. Penguin, Harmondsworth & New York.

Hamilton, W. D., & Zuk, M. (1982). Heritable true fitness and bright birds: A role for parasites? *Science* 218, 384–387.

Jones, D. (1996). An evolutionary perspective on physical attractiveness. *Evolutionary Anthropology* 5, 97–109.

Lévi-Strauss, Claude (1963). *Structural Anthropology*. Basic Books, London.

Lewin, Roger (1997). *Patterns in Evolution: The New Molecular View*. Scientific American Library, New York.

Møller, A. P., & Swaddle, J. P. (1997). *Asymmetry, Developmental Stability and Evolution*. Oxford University Press, Oxford.

Peel, Robert A. (1997). *Marie Stopes, Eugenics and the English Birth Control Movement*. The Galton Institute, London.

Russell Jones, R., & Southwood, R. (1987). *Radiation and Health: The Biological Effects of Low-level Exposure to Ionizing Radiation*. John Wiley & Sons, Chichester & New York.

Schievenhövel, W. (1989). Reproduction and sex ratio manipulation through preferential female infanticide among the Eipo in the Highlands of West New Guinea. In Rasa, A. E., Vogel, E. C., & Voland, E. (eds.), *The Sociobiology of Sexual and Reproductive Strategies*, pp. 170–193. Chapman & Hall, London.

Short, R. V. (1976). The evolution of human reproduction. *Proceedings of the Royal Society of London* 195, 3–24.

Stopes, M. C. (1918). *Married Love: A New Contribution to the Solution of Sex Difficulties*. A. C. Fifield, London.

Trivers, R. L., & Willard, D. E. (1973). Natural selection of parental ability to vary the sex ratio of offspring. *Science* 179, 90–92.

Williams, G. C. (1966). *Adaptation and Natural Selection: A Critique of Some Current Evolutionary Thought*. Princeton University Press Princeton, N. J.

Chapter 4 Playing God

Bianchi, D. W. (1995). Prenatal diagnosis by analysis of fetal cell in maternal blood. *Journal of Pediatrics* 127, 847–856.

Brock, D. J. H., Rodeck, C. H., & Ferguson-Smith, M. A. (1992). *Prenatal Diagnosis and Screening*. Churchill Livingstone, Edinburgh.

Clarke, Angus (1994). *Genetic Counselling: Practice and Principles*. Routledge, London.

Coutelle, C., Douar, A.-M., Colledge, W. H., & Froster, U. (1995). The challenge of fetal gene therapy. *Nature Medicine* 1, 864–866.

Couture, L. A., & Stinchcomb, D. T. (1996). Anti-gene therapy: The use of ribozymes to inhibit gene function. *Trends in Genetics* 12, 510–515.

Cuckle, H. (1996). Established markers in second trimester maternal serum. *Early Human Development* 47 Supplement: S27–S29.

Dale, B., & Elder, K. (1997). *In Vitro Fertilisation*. Cambridge University Press, Cambridge.

Fiddler, M., & Pergament, E. (1996). Germline gene therapy: The time is near. *Molecular Human Reproduction* 2, 75–76.

Finger, Anne (1991). *Past Due—A Story of Disability, Pregnancy and Birth*. The Women's Press, London.

Hamer, D. H., Hu, S., Magnusson, V. L., Hu, N., & Pattatuci, A. M. L. (1993). A linkage between DNA markers in the X chromosome and male sexual orientation. *Science* 261, 321–327.

Handyside, A. H., Kontogiani, E. H., Hardy, K., & Winston, R. M. L. (1990). Pregnancies from biopsied human preimplantation embryos sexed by Y-specific DNA amplification. *Nature* 244, 768–770.

Highfield, Roger. Should this child have been born? *The Daily Telegraph*, September 17, 1997, p. 16.

Howell, M., & Ford, P. (1980). *The History of the Elephant Man.* Penguin, Harmondsworth.

Jones, Steve (1996). *In the Blood.* HarperCollins, London.

McGee, Glenn (1997). *The Perfect Baby—A Pragmatic Approach to Genetics.* Rowman & Littlefield, Lanham, Md., & London.

McKusick, Victor (1983). *Mendelian Inheritance in Man.* Johns Hopkins University Press, Baltimore.

Modell, B., & Modell, M. (1992). *Towards a Healthy Baby.* Oxford University Press, Oxford.

Moore, Pete (1997). *Pregnancy—A Testing Time.* Lion Press, Oxford.

Mueller, R. F. (1998). *Emery's Elements of Medical Genetics*, Tenth Edition. Churchill Livingstone, Edinburgh.

O'Connor, N., & Hermelin, B. (1988). Low intelligence and special abilities. *Journal of Child Psychology and Psychiatry* 29, 391–396.

Ohlsson, R., Hall, K., & Ritzen, M. (1995). *Genomic Imprinting: Causes and Consequences.* Cambridge University Press, Cambridge.

Plomin, R., DeFries, J. C., McClearn, G. E., & Rutter, M. (1997). *Behavioral Genetics*, Third Edition. W. H. Freeman, New York.

Schnieke, A. E., Kind, A. J., Ritchie, W. A., Scott, A. R., Ritchie M., Wilmut, I., Colman, A., & Campbell, K. H. S. (1997). Human factor IX transgenic sheep produced by transfer of nuclei from transfected fetal fibroblasts. *Science* 278, 2130–2133.

Thornton, J. G. (1994). Prenatal screening programmes. *Lancet* ii, 1090–1091.

Wald, N. J. (ed.) (1984). *Antenatal and Neonatal Screening.* Oxford University Press, Oxford.

Weatherall, D. J. (1991). *The New Genetics and Clinical Practice*, Volume 3. Oxford University Press, Oxford.

Chapter 5 Keep Out the Clones

Briggs, R., & King, T. J. (1952). Transplantation of living nuclei from blastula cells into enucleated frogs' eggs. *Proceedings of the National Academy of Sciences of the USA* 38, 455–463.

Bulmer, M. G. (1970). *The Biology of Twinning in Man.* Clarendon Press, Oxford.

Bryan, Elizabeth A. (1983). *The Nature and Nurture of Twins.* Ballière Tindall, London.

Campbell, K. H. S., McWhir, J., Ritchie, W. A., & Wilmut, I. (1996). Sheep cloned by nuclear transfer from a cultured cell line. *Nature* 380, 64–66.

Dafoe, A. R. (1934). The Dionne quintuplets. *Journal of the American Medical Association* 03, 673–677.

Dessauer, H. C., & Cole, C. J. (1986). Clonal inheritance in parthenogenetic whiptail lizards: Biochemical evidence. *Journal of Heredity* 77, 8–12.

Gurdon, J. B., & Uehlinger, V. (1966). "Fertile" intestinal nuclei. *Nature* 210, 1240–1241.

Kluger, Jeffrey. Will we follow the sheep? *Time*, March 10, 1997.

Kolata, Gina (1998). *Clone: The Road to Dolly and the Road Ahead.* William Morrow, New York.

Meng, L., Ely, J. J., Stouffer, R. L., & Wolf, D. P. (1997). Rhesus monkeys produced by nuclear transfer. *Biology of Reproduction* 57, 454–459.

Newman, H. H., & Patterson, J. T. (1910). The development of the nine-banded armadillo from the primitive streak stage to birth; with especial reference to the question of specific polyembryony. *Journal of Morphology* 21, 359–423.

Steward, F. C., Mapes, M. O., Kent, A. E., & Holsten, R. D. (1964). Growth and development of cultured plant cells. *Science* 143, 20–27.

Willadsen, S. M. (1986). Nuclear transplantation in sheep embryos. *Nature* 320, 63–65.

Wilmut, I., Schnieke, A. E. McWhir, J., Kind, A. J., & Campbell, K. H. S. (1997). Viable offspring derived from fetal and adult mammalian cells. *Nature* 385, 810–813.

Chapter 6   Sex Selection

Beernink, F. J., Dmowski, W. P., & Ericsson, R. J. (1993). Sex preselection through albumin separation of sperm. *Fertility and Sterility* 59, 382–386.

Carson, S. A. (1988). Sex selection: the ultimate family planning. *Fertility and Sterility* 50, 16–19.

Cran, D. G., & Johnson, L. A. (1996). The predetermination of embryonic sex using flow cytometrically separated X and Y spermatozoa. *Human Reproduction Update* 2, 355–363.

Ericsson, R. J., Langevin, C. N., & Nishino, M. (1973). Isolation of fractions rich in human Y sperm. *Nature* 246, 421–424.

Flaherty, S. P., Michalowska, J., Swann, N. J., Dmowski, W. P., Matthews, C. D., & Aitken, R. J. (1997). Albumin gradients do not enrich Y-bearing human spermatozoa. *Human Reproduction* 12, 938–942.

Fugger, E. F., Black, S. H., Keyvanfar, K., & Schulman, J. D. (1998). Births of normal daughters after MicroSort sperm separation and intrauterine insemination, in-vitro fertilization, or intracytoplasmic sperm injection. *Human Reproduction* 13, 2367–2370.

Johnson, L. A. (1994). Isolation of X- and Y-bearing sperm for sex preselection. *Oxford Reviews of Reproductive Biology* 16, 303–326.

Human Fertilisation and Embryology Authority (1993). *Sex Selection*. Public Consultation Document, Her Majesty's Stationery Office, London.

Johnson, L. A., Flook, J. P., & Hawk, H. W. (1989). Sex preselection in rabbits: Live births from X and Y sperm separated by DNA and cell sorting. *Biology of Reproduction* 41, 199–203.

Lindsey, J. K., & Altham, P. M. E. (1998). Analysis of the human sex ratio by using overdispersion models. *Applied Statistics* 47, 149–157.

Martin, R. (1994). Human sex preselection by sperm manipulation. *Human Reproduction* 9, 1790–1791.

Moysa, Marilyn. A boy or girl? Answer can be terminal. *The Edmonton Journal*, October 12, 1996, p. 1.

Pyrzak, P. (1994). Separation of X- and Y-bearing human spermatozoa using albumin gradients. *Human Reproduction* 9, 1788–1790.

Statham, H., Green, J., Snowdon, C., & France-Dawson, M. (1993). Choice of baby's sex. *Lancet* 341, 564–565.

Chapter 7   Other Wombs

Barker, D. J. P., Winter, P. D., Osmond, C., Margetts, B., & Simmonds, S. J. (1989). Weight in infancy and death from ischaemic heart disease. *Lancet* 2, 577–580.

England, Marjorie A. (1983). *A Colour Atlas of Life before Birth: Normal Fetal Development*. Wolfe Medical Publications, Weert, Netherlands.

Gill, T.J., III (1994). Reproductive immunology and immunogenetics. In Knobil, E., & Neill, J. D. (eds.), *The Physiology of Reproduction*, Second Edition, Vol. 2. Raven, New York.

Selwood, L., & Woolley, P.A. (1991). A timetable of embryonic development, and ovarian and uterine changes during pregnancy, in the stripe-faced dunnart, *Sminthopsis macroura* (Marsupialia: Dasyuridae). *Journal of Reproduction and Fertility* 91, 213–227.

Stabile, Isabel (1996). *Ectopic Pregnancy: Diagnosis and Management*. Cambridge University Press, Cambridge.

Unno, N., Kuwabara, Y., Okai, T., et al. (1993). Development of an artificial placenta: Survival of isolated goat fetuses from three weeks with umbilical arteriovenous extracorporeal membrane oxygenation. *Artificial Organs* 17, 996–1003.

Chapter 8   Never Too Late?

Austad, S. N. (1997). *Why We Age*. John Wiley & Sons, New York.

Berryman, J., Thorpe, K., & Windridge, K. (1995). *Older Mothers: Conception, Pregnancy and Birth after 35*. Pandora, London.

Cummins, J. M., Jequier, A. M., & Kan, R. (1994). Molecular biology of human male infertility: Links with aging, mitochondrial genetics, and oxidative stress? *Molecular Reproduction and Development* 37, 345–362.

Fédération CECOS, Schwartz, D., & Mayoux, N. J. (1982). Female fecundity as a function of age: Results of artificial insemination in 2193 nulliparous women with azoospermic husbands. *New England Journal of Medicine* 306, 404–406.

Finch, C. E. (1990). *Longevity, Senescence and the Genome.* University of Chicago Press, Chicago

Gosden, Roger (1996). *Cheating Time: Science, Sex and Aging.* W. H. Freeman, New York.

Gosden, R. G., & Aubard, Y. (1996). *Transplantation of ovarian and testicular tissues.* R. G. Landes, Austin, Tex.

Gosden, R. G., Baird, D. T., Wade, J. C., & Webb, R. (1994). Restoration of fertility to oophorectomised sheep by ovarian autografts stored at –196°C. *Human Reproduction* 9, 597–603.

Gow, S. M., Turner, E. I., & Glasier, A. (1994). Clinical biochemistry of the menopause and hormone replacement therapy. *Annals of Clinical Biochemistry* 31, 509–528.

Greer, Germaine (1991). *The Change: Women, Aging and the Menopause.* Penguin, Harmondsworth.

Meldrum, D. R. (1993). Female reproductive aging-Ovarian and uterine factors. *Fertility and Sterility* 59, 1–5.

Menken, J., Trussell, J., & Larsen, U. (1986). Age and infertility. *Science* 233, 1389–1394.

Navot, D., Bergh, P. A., Williams, M. A., Garrisi, G. J., Guzman, I., Sandler, B., & Grunfeld, L. (1991). Poor oocyte quality rather than implantation failure as a cause of age-related decline in female fertility. *Lancet* 337, 1375–1377.

Paulson, R. J., & Sauer, M. V. (1994). Pregnancies in post-menopausal women. *Human Reproduction* 9, 571–572.

Chapter 9 Reproductive Liberties

Blood, D. (1998). Response to the consultation document of Professor S. McLean. *Human Reproduction* 13, 2654–2656.

de Bono, Edward (1998). *Simplicity.* Viking, London.

Caplan, Arthur (1998). *Am I My Brother's Keeper? The Ethical Frontiers of Biomedicine.* Indiana University Press, Bloomington.

Council of Europe (1997). *Convention on Human Rights and Biomedicine.* Strasbourg, April 1997.

Dworkin, Ronald (1993). *Life's Dominion—An Argument about Abortion and Euthanasia.* HarperCollins, London.

Edwards, Robert (1989). *Life Before Birth: Reflections on the Embryo Debate.* Hutchinson, London.

Harris, John (1992). *Wonderwoman and Superman: The Ethics of Human Biotechnology.* Oxford University Press, Oxford.

Harris, John, & Holm, Søren (eds.) (1998). *The Future of Human Reproduction: Ethics, Choice and Regulation.* Oxford Clarendon Press, Oxford.

Marteau, T., and Richards, M. (1996). *The Troubled Helix: Social and Psychological Implications of the New Human Genetics.* Cambridge University Press, Cambridge.

Mead, Margaret (1950). *Male and Female: A Study of the Sexes in a Changing World.* Victor Gollancz, London.

Nesse, R. M., & Williams, G. C. (1996). *Why We Get Sick—The New Science of Darwinian Medicine.* Vintage Books, New York.

Robertson, John A. (1994). *Children of Choice: Freedom and the New Reproductive Technologies.* Princeton University Press, Princeton, N.J.

Suzuki, David, & Knudtson, Peter (1990). *Genethics.* Harvard University Press, Cambridge, Mass.

Vaughan, A. T., & Mason Vaughan, V. (1991). *Shakespeare's Caliban: A Cultural History.* Cambridge University Press, Cambridge, New York, & Melbourne.

Winston, Robert (1987). *What We Know about Infertility: Diagnosis and Treatment Alternatives.* Free Press, New York.

Index